From Snowshoes to Wingtips

The Life of Patrick O'Neill

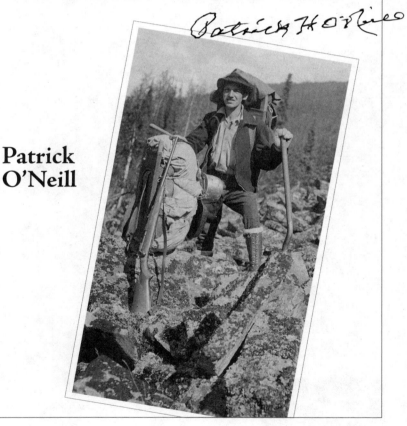

Patrick O'Neill

University of Alaska Foundation
Fairbanks

Distributed by
University of Alaska Press
P.O. Box 756240
Fairbanks, AK 99775-6240

ISBN 10: 1-883309-05-0
ISBN 13: 978-1-883309-05-3

Cover image: Patrick prospecting in the Goodpaster area in Alaska, 1935.

UNIVERSITY *of* ALASKA
FOUNDATION

To the love of my life, my wife Sandra.

Without her this book would not have been written.

Table of Contents

Foreword

ALTHOUGH I have respected and treasured the friendship of Pat O'Neill for innumerable years and have been well aware of his reputation in the mining world, it was not until I read his memoir that I fully comprehended the extent of the challenges he underwent during a long and incredibly successful career.

Born in 1915 in Cordova, Alaska, the seventh of 12 children, Pat's career began as a young teenager working ten hours a day, seven days a week at a mine in Alaska, living in a bunk in a tent with "three good meals and collecting $5.00 a day."

Working his way through the University of Alaska Fairbanks over an eight-year period, he earned two degrees in mining engineering before war service as a pilot between 1941 and 1946.

Returning to his mining career, Pat describes how he coped as a manager of a gold and platinum dredging operation in Colombia, South America, with problems created by illiteracy, poverty-level wages, inadequate housing, and poor medical attention, which resulted in his company sustaining huge losses. The Herculean measures he took to successfully resolve these problems launched Pat on a path in his profession greatly admired by all who have known him.

Holding executive positions over the years with a number of mining companies, he faced many complex, frequently life-threatening challenges, particularly in Colombia, in an effort to operate profitably under difficult conditions with the constant threat of expropriation looming.

Serving as a fellow director with Pat for 28 years on the board of Zemex Corporation gave me sufficient time to take the measure of a man of great integrity, possessing formidable leadership qualities, who just happens to know how to write about his unique career in a most engaging and engrossing manner. This is a "must-read" memoir.

—*Peter Lawson-Johnston*

Acknowledgments

FIFTEEN years ago when I asked my wife, Sandra, what she would like for Christmas, she said that she would like me to write my memoirs, which I promised to do. I started out with paper and pencil and wrote when I had spare time, usually on cruises. I learned about word processing and put my work on the computer from time to time, but I didn't work diligently until my late 80s, when I realized that I had best get busy if I ever expected to finish. I am indebted to Sandra, and to our children, Kevin and Erin, for their encouragement, constructive criticism, and suggestions throughout this long process.

My first job in mining was working under my brother, Bill, who taught me to work as hard as I could. Then I worked under Jim Crawford in Fairbanks, who taught me to work intelligently as well as hard in engineering and organization. I was concerned that I had made a mistake when I went to Colombia, but Lew Harder arrived at a critical time and took me to New York for a long and exciting career in the management of several international companies. I was fortunate also in serving on the boards of many corporations with outstanding executives over the years. During 28 years on the board of Pacific Tin, later Zemex, the chairman during nearly the entire time was Peter Lawson-Johnston, whom I admired greatly for his calm, perceptive, and consistent manner with the board. I am indebted to Peter for writing the foreword to this book.

I am also indebted to the editor, Erica Hill, for her advice, patience and guidance in putting this book together, and to MaryLou Wilkinson, who has been a great help with the computer.

1 Cordova and Family

THE brother just older than me, Mike, was the sixth in our family. My parents had decided to call him Michael Patrick, but as they were on their way to church to baptize him my dad said, "It's a shame to waste two good names on one kid—let's call this one Michael and save Patrick for the next one." I have always been very happy with that decision as I have sincerely liked my name, Patrick, and have been very proud of it.

My paternal grandfather, William A. O'Neill, was born in Liverpool in 1855. His parents were from Dundalk in County Louth in Ireland but moved to Liverpool during the famine, and Grandpa was born there soon after they arrived. Grandpa immigrated to America in 1871 when he was 16, and worked his way west as a laborer on railroad construction. When the crew was stationed near Fargo, North Dakota, he married, became a farmer, and raised ten children. One of these was my father, Harry Irenaeus O'Neill, who was born on June 28, 1885. At the time of the 1897 Klondike gold rush Grandpa went north to Alaska and the Yukon. He carried his supplies up the famous Chilkoot Trail from Dyea, near Skagway, to a mountain pass that lay on the border with Canada. The North West Mounted Police required a year's supply of food, roughly 1,000 pounds, for each person entering Canada; each man therefore had to make numerous trips to get his food and supplies to the top and then had many miles to transport it to the headwaters of the Yukon River, where they built boats or rafts to go down the river to Dawson. There was a constant line of men carrying their supplies up the trail, and one day there was a snowslide that buried many men. Grandpa fortunately missed the slide and helped dig out many of the men and bury them properly. He did not succeed in finding the riches dreamed of by so many in those days, so he went back to railroad construction on the White Pass and Yukon Railroad. The construction started in Skagway, Alaska, in May 1898 and terminated in Whitehorse, Yukon Territory, on July 29, 1900. He then went back to North Dakota

Map of Alaska showing the location of Cordova and Fairbanks.

but moved his family to Seattle in 1901 and returned to the Yukon; he worked at various jobs, including operating a roadhouse, until he went back to railroad construction. The builder of the White Pass and Yukon Railroad contracted with the Guggenheim interests to build a railroad from Eyak (later named Cordova) to Kennicott, to open up the great copper mines that were the start of the Kennecott Copper Corporation. (The *i* in the name of the early explorer Robert Kennicott was inadvertently changed to an *e* when the company was incorporated.) Known as the Copper River and Northwestern Railroad, it was built between 1905 and 1911. Grandpa O'Neill was a foreman on that construction job from the start until he was killed in August 1911 in a cave-in of the last tunnel near Chitina, close to the end of the railroad construction.

The parents of my maternal grandfather, Philip J. Leahy, came to America from County Clare in Ireland, and Grandpa Leahy was born in Holmewood, Pennsylvania, in 1858. He worked his way west as a telegrapher on the central railway system at about the same time that Grandpa

O'Neill was working on the northern system. On November 14, 1883, while he was stationed there, my mother, Florence Anne Leahy, was born in Beardstown, Illinois. Grandpa Leahy and his family ended up in Seattle at about the same time that Grandpa O'Neill moved his family there, and my parents met while they were both working in a grocery store. When they married in 1904, Mother was 21 and Dad was 18. Neither of them had gone very far in school. Mother had three younger brothers, and her mother died not long after the youngest one was born, so Mother left school early to help at home. However, she always had us taking courses in school that she wanted to learn. She studied Latin with me, among other things; at the time I wished she could take the test for me as she was better at it than I was. Dad, too, had to cut short schooling and work on the farm because when he was 12 his father left for the Klondike gold rush. However, Dad did very well in business later on.

The name of the town of Cordova was taken from nearby Bahia de Cordoba, founded in 1790 by a Spanish explorer, Salvador Fidalgo, and named after Spanish officer Luis de Cordoba. Grandpa O'Neill acquired a building lot for my parents while he was working on the layout of the town.

Town of Cordova, Alaska, where Patrick was born and raised.

Family store, Cordova, 1930.

By the time they had the wherewithal to travel north in 1908, they had two children, and ten more followed with regularity after their arrival in Cordova. At first they lived in a small shack as Cordova was mostly a tent town at that time, but later they built a large house on the lot Grandpa had acquired for them. I was born on August 11, 1915, as the seventh of seven boys and five girls. Unfortunately, Grandpa had been killed in 1911 so I never had the opportunity to know him.

Dad worked for Sam Blum in a tent store that expanded frequently. Later, it became Blum and O'Neill, and when Sam died, Dad bought out Mrs. Blum and it became the O'Neill Co. Inc., which by then had three separate departments: groceries, hardware, and clothing. We all worked in the store after school and during vacations, from the time we could stock shelves, count items for inventory, or wield a broom or shovel.

At home we were organized into pairs for household chores. We started off taking out ashes, bringing in kindling and coal, and stoking the cookstove and the furnace; we then progressed to sweeping and scrubbing floors, dishwashing, ironing clothes, and helping with the cooking. In addition to our being regular altar boys on Sundays Mother also liked to volunteer our services on weekdays, so we took turns at early Mass, summer and winter. Our house was one of the largest, if not the largest, in town. There were three stories, including a full basement that was not finished off but which had the furnace and laundry facilities. The main floor had a large kitchen with a seating area for about six, a large dining area, a large living room, five bedrooms, and one bathroom. The girls slept in bedrooms, and the oldest boys also had a bedroom. Five of us boys slept in the attic, which also had a large area of clotheslines for drying the laundry. The washing machine in the basement had a wringer but no dryer, and Mother carried all the laundry up two flights of stairs to hang everything in the attic to dry. We all helped when we were home, but she did most of the laundry while we were in school and working in the store. I always marveled at all she did, and we each had clean clothes every day.

One time Father Mac asked Mike and me to tap a new keg of wine for him, which we did. We thought we should taste it, though, to make sure that it was all right, and Father Mac caught us. He said that there was a big difference between tasting and drinking and was really mad at us. We thought (hoped) that we might get fired as altar boys, but Mother was informed and she chastised us and then signed us up to serve the daily Mass for our entire school vacation. One picture that Mother was very proud of was of all seven of us boys on the altar at the same time as servers for a Christmas Mass.

During the week we all ate in shifts in the large kitchen, but every Sunday we all ate together in the dining room, which could sit up to 24 comfort-

Family home in Cordova, where Patrick was raised with eleven brothers and sisters.

ably. Almost every Sunday we had company and often had 20 or more for Sunday dinner. This was frequently leg of lamb, although on more informal occasions we also had ham and cabbage, which I always liked, especially the hash with the leftovers the next day. After dinner on most Sundays the family and guests would walk downtown to the movie theater, except for the two who were on kitchen detail at the time. One Sunday when Mike and I were doing the dishes I was to do the scraping and rinsing, and then Mike was to do the washing while I dried. Mike went outdoors, so when I had things ready I went out to get him. He did not answer my call, so I got on a pogo stick and went bouncing up and down alongside the house yelling for him. I turned my head when I thought I heard him and landed on a rock, going headfirst through a window and gashing my nose. It was bleeding profusely, and we kept putting cold cloths on it but could not stop the bleeding, so we walked to the hospital with extra cold cloths that we changed frequently. The only doctor in town lived in the hospital, but he

Patrick making sure everyone is in line.

Patrick in front with seven of his brothers and sisters.

was out on a drinking binge, so the nurse sent Mike to the movie theater to get Dad to go and look for him. Dad did find him, but he was quite drunk. Dad and the nurse sobered him up somewhat with coffee, and he finally stitched up my nose as well as he could, but he was weaving around while the two of them tried to help him and hold him steady. I ended up with a very irregular scar on my nose that was quite noticeable for many years, although now almost 80 years later it is hardly visible.

We all had to get dressed up for Sunday dinner, and the smaller boys each had one white outfit to wear. Mother would let us go out to play while waiting for dinner, with dire threats if we got dirty. Cows often grazed in the field alongside our house, and one time when we were out Edward pushed me and I fell into a cow pile; not only did I get spanked but, not having any other nice clothes, I could not eat except in the kitchen, after everyone else was finished.

Mother was a very small person, just barely five feet tall; Dad was just over six feet. None of the 12 of us ever got as tall as Dad but all of us passed Mother, although two of the younger girls by just a little bit. It was interesting watching them walk together: Mother would walk one step and run two or three to keep up with Dad's long stride. Mother spoke softly but had a big stick in the closet that she seldom had to use. Just starting to go to the closet was enough for us to quickly decide that we must do as we were told. Before we were big enough to work in the store, we had jobs such as selling or delivering newspapers, baby-sitting, herding cows in the summer, delivering chickens for a neighbor who raised them, and the like. As long as we lived at home we gave Mother half of our earnings—on such jobs and also later when we worked in the store or elsewhere.

Mike and I were partners on the house chores, and we worked together on some of these other jobs as well. There was no kindergarten in Cordova and one had to be six before the end of December to start first grade. Mike's birthday is in January, so he was over six and a half by the time he was allowed to start school. I had turned five in August and was the same size as Mike, and after a great deal of wheedling we were dressed as twins and sent off to school together. Mother filled out Mike's information card completely and only put my name on mine and clipped it to Mike's card; this was as far as she would go in implying that we were twins, but she advised us that if anyone asked directly we must tell the truth. Fortunately, no

Mike and Pat dressed as twins with Rusty and Francis.

one asked until we were in the third grade, so they let me stay and we went all through school together—although not always very peacefully, as we fought frequently and Mike regularly bested me. At long last when I was 15 I fibbed by three years and got a summer job in a mine. With every shovel of dirt, and every time I wielded pick or axe, I flexed what few muscles I had and thought, wait until I get Mike. Finally I arrived home on the train and was walking up the road to our house with my bedroll and duffel bag on my back. One of my younger sisters saw me and alerted Mother, so there was quite a reception on the porch, but just as I approached the stairs Mike came around the corner of the house, so I dropped my bags and went at him. He had been working as well, so the battle wasn't as easy as I had expected, and we had a tough fight that ended in a draw. That was the last fight we ever had and we were good friends ever after. Mother's only comment was, "At least you should have first come to say hello to me."

Mike and I acquired a young goat somehow and we had a great time with it for awhile. Mike would imitate the goat with its bleating, and he still sounds somewhat like the goat when he laughs. The goat would get away from us occasionally and liked to get into neighbors' homes and leave its calling card, so we were frequently running around with a broom, dustpan, and mop cleaning up after the goat. One time the goat got into our storage pantry and ate the labels off many of the cans, so Mother wasn't certain for quite awhile whether she was opening a can of beans or peas or whatever. Then one day the goat butted Dad when he stooped to pick up something, so we had to give it away. We gave it to a friend of Dad's who lived on an island not far from Cordova. The man set the goat on the dock, but when he stooped over to tie his boat, the goat butted him into the bay, so he had it for dinner the next night.

We were young during Prohibition, and the bootleggers would have their stills out on one of the islands. They would bring their kegs of moonshine ashore at night by rowboat and cache them, then go back to their boat and motor over to tie up at the dock, where the Prohibition agents were always on the lookout. The bootleggers would then go into town to negotiate the sale and tell the buyer where the whiskey was. Mike and I would occasionally walk along the shore after school with a wheelbarrow, and once in awhile we would find a keg, cover it with old gunnysacks, and take it home and hide it in the coal bin. We would tap the keg and siphon off some of the whiskey into a flask, which we would then take to a speakeasy as a sample. If we could negotiate a sale, we would take the keg the next night to the place designated by the buyer. One time when we were out in the woods near our house we found a small keg, which we took home and tapped. It was the best whiskey we ever found, and we negotiated a good price. Many years later at a family reunion in Anchorage we were recounting some of the escapades of our youth and Mike mentioned this whiskey, and Dad became very agitated and said, "So you are the ones that stole my whiskey. If I had caught you then, I would have given you a real beating; that was exceptionally good whiskey and I'm sure I paid a lot more for it than you stupid kids sold it for." Which was undoubtedly true: we had acquired enough of a taste to know good whiskey when we found it, but we were certainly naive about the value.

During Prohibition many people made beer (home brew) or wine. Mother decided that she would make beer, as she said that we were going to drink anyway and she would rather we did it at home instead of running off to someone else's house to drink. Her beer was so good that we had many friends coming to our house, and I think she started regretting making beer. One of Mother's little quotes that I always liked was, "I know my boy doesn't drink at night; he is always so thirsty in the morning."

One of my earlier jobs was delivering newspapers. One day I was rolling up a paper to wedge it behind a doorknob when a big black dog came around the corner and attacked me. It took a nip on the upper part of my leg and then bit a chunk out just above my left ankle. I got away from it and ran towards home, about a mile away, very foolishly passing not far from the hospital en route. Mother put a tourniquet above the wound and sent me to the hospital, where they sewed up the wound. However, infection set in and gangrene developed, so I was back in the hospital, where they applied hot compresses. After two days the doctor told Dad that if there wasn't any improvement soon, he would have to amputate the lower part of the leg. I told Dad that the day nurses were changing the compresses often but the night nurse was not, as her boyfriend was spending most of the night with her. Dad stayed with me for several nights to be sure that the compresses were being changed, and the leg soon got better. I went back to my paper route, but shortly after that a dog came after me and bit me on the rear, so I stopped the delivery route and went over to selling papers on the street.

There were only seven miles of road in Cordova along Lake Eyak, and a big outing for us would be to get a ride out there in the back of a truck after big events like First Communion or other celebrations and have a picnic. Mike and I would occasionally hike out along the railroad tracks to visit the camps at Mile 13 or 18, where the men were cutting ties for the railroad. One time we were walking in the woods close to where one of the men was working when we encountered a bear. We turned and ran all the way back to the camp without ever looking back. We were really scared. The man told us later that he had seen us and that the bear had turned and run just as fast away from us and that he had a good laugh watching the encounter. Sometimes we would walk to Mile 22, Alaganik, where a bootlegger named McAllister had a small roadhouse on the bank of a branch of the Copper River close to the railroad. Mac was a great storyteller and a wonderful cook. He would make the best pancakes and flip them to whoever was ready at a table eight or ten feet away from the stove, so you had to be a fair catcher to get your share of pancakes. Mac would make us eat garlic sandwiches in the evening to prevent catching cold. One time he took us out duck hunting in his small boat. He let me use a double-barreled shotgun and cautioned me about the force if both barrels were fired at the same time. When some ducks suddenly appeared I got excited and pulled both triggers and was knocked overboard. Mac wasn't very concerned about me nearly drowning, but he was furious about his gun getting wet. Another old-timer named Scotty Malone lived a few miles back in the woods and we used to visit him as well. There was an old mine there and Scotty said that he was a prospector, but we were sure that he was also making moonshine. Scotty was the dirtiest man I ever saw and you could smell him yards away. He never bathed, and when one suit of underwear wore out he would put a new one on over the old. He became ill and was taken to the hospital in Cordova where they stripped him of his filthy clothes and bathed him, and the poor guy died very soon thereafter. I always thought it was from exposure and shock.

The Fish and Wildlife Service agent in Cordova was transferred shortly after obtaining a pair of male and female Newfoundland dogs. He gave them to me before he left, saying that they would produce pups that I could sell for a good price, sending him 50 percent of what I made. Dad helped build a fenced area and doghouses close to our house. I was told that I had to feed them canned dog food, but they were voracious eaters and I soon ran out of money, so I arranged to pick up scraps daily from the local restaurants. We did everything we could to encourage the dogs to breed but without success, so after a year I sold the male and took the female out to McAllister at Alaganik on one of our trips out there. Mac liked the dog and asked to keep her, so I left her with him. She was running loose and it wasn't long before she bred with a wolf. She had one big pup and died soon after. The half-wolf pup grew to be a very large and strong animal. Mac trained it to pull a sled, and it could pull a very big load for hours all by itself. Mac used it to haul supplies to his still and to take the moonshine to Cordova, as well as for hunting and trapping.

The summer I was seven some friends and I made rafts from old railroad ties to play with in a slough down by the railroad yards. We were rafting after school on the day that Mother went to the hospital for the birth of her 12th child, my youngest sister Rose Mary. While we were changing rafts, the one I was to jump onto was pushed out when my friend jumped to mine, so I missed his when I jumped. I didn't know how to swim, and as I was going down for the third time the last thing I remembered was thinking what a terrible thing for my mother to have one child drown while she was giving birth to another. Fortunately my friends pulled me out and revived me, but it was quite awhile before Mother learned about my escapade.

In 1923 President Harding visited Alaska and spent a day in Cordova. I was in grade school at the time, and the entire student body was lined up along his route from the dock to town waving flags and singing the Alaska song. A platform had been built on Main Street so that people could walk up one side, shake the president's hand, and continue down the other side. A very big fellow, Babe Hayes, was the head of the chamber of commerce and he was introducing everyone. Our family was all lined up in order and as we approached the stand Babe said, "Mr. President, here are Harry and Florence O'Neill and their 12 children." I recall the president looking down the line and saying, "Well, let's get on with it," and we all filed past shaking his hand on the way.

In one of the lower grades we had a large, tough teacher who read a story to us each afternoon and while reading would not recognize anyone who had to go to the bathroom. We had combination seats and desks, with the seats curved across. Woody and I sat behind Elmer, who would often wet his pants while the teacher was reading and ignoring his raised hand. Woody and I would wager as to which side the urine would run out that day, and whenever I won I could not help chuckling. The teacher would stop reading, come to my desk, and rap my knuckles with a ruler. If I did it two days in a row, she would sit me in the corner with a dunce cap. That cap must have been made for my larger-than-average head as I used it more often than anyone else in the class.

We went through all 12 grades in the one building. The classes were all small, some as small as two or three students. My class, which graduated eight, was the largest up to that time. Many years later I took my family to see Cordova, which was still a small town of 2,500 or thereabouts. When I showed them the school I had attended, my daughter asked which grades I had gone through there. When I told her all 12 she was very surprised and said, "Gee, Dad, it isn't as big as our home in Connecticut." Although the school was very small, it must have been reasonably good, as four out of the seven or eight of us graduates that went on to the University of Alaska earned the distinction of Distinguished Alumnus of the University. In addition to me these included my older brother William, Bettie Harrop Clark, and Woodrow Johansen.

Cordova had very few cars and no school buses or any other types of buses, so everyone walked wherever they had to go. Our home was near the base of a steep hill that we had to climb to go to school and also to church. Climbing the hill was difficult, especially in winter. The hillside was littered with logging debris and stumps through which we had to pick a path. When there was snow and ice on the ground, which was usual for at least five months of the year, the trail was slippery and hard to climb. I vividly remember my mother struggling up the hill to go to church. She had "creepers," which helped; these were metal pieces with sharp points that she would pull over the front of her galoshes. When we were in the seventh grade or thereabouts I damaged one knee and was on crutches for a month or more. I could not climb the hill on crutches so would get out early and wait for the group that came by our house from Eyak, or "old town," as it was called. The biggest one in our class was Woody Johansen, and he would let me jump on his back and be carried up the hill. Several years after I left Cordova I was glad to see on a visit that the city had built stairs up the hill, but I never understood why it had not been done many years earlier. Woody was not only the biggest and strongest but also the best student and the nicest person in our class. I mention him elsewhere in these memoirs as the one that we all tied onto as he led us through deep snowdrifts in a blizzard to the dock when we were on a basketball trip to Anchorage. He was also the one who graduated first in the class, as I describe in the passage where I refer to myself being eighth.

Cordova once took a leaf from the history of Boston and the famous tea party. In 1906 the U.S. government withdrew from public entry the coal, oil, and timberlands of the Territory of Alaska. Coal was being mined at Katalla, a few miles from Cordova, where the first attempts to build a railroad to the copper mines had been started but then abandoned after a storm destroyed the docks and other facilities. Without the coal from Katalla the pioneering citizens were compelled to purchase, at exorbitant prices, Canadian coal shipped over a thousand miles. Protests to the government were unheeded, and Washington turned a deaf ear to the pleas. Indignation and the sense of injustice grew until the citizens reached the point of desperation. Then one day when a shipment of Canadian coal had just been unloaded three hundred Cordova citizens armed with shovels marched to the dock and, to the cry, "Give us Alaska coal," proceeded to pitch coal into the bay. It was not a hoodlum mob, as it was led by the president of the chamber of commerce and included most of the business and professional men of the town. The U.S. deputy marshal was conveniently out of town, and by the time the chief of police obtained a federal warrant several tons of coal had been thrown into the bay and the demonstration had served its purpose. News flashes of the "Coal Party" were spread throughout the country and assurance soon came from Washington that the land laws would be modified, which they were in due course.

The Copper River and Northwestern Railroad was the essential link in moving the copper ore from the mines at Kennicott to Cordova, where it was shipped on the weekly steamer to the smelter at Tacoma, Washington. The snowfall was very heavy, requiring large snowplows in the winter, and in some areas the railroad was constructed on glaciers and required constant repair and realignment. At Chitina the railroad crossed the Copper River, where the spring floods usually washed out the bridge and tracks. At the time, builders decided that it would be cheaper to replace a wooden bridge every year than to build a more substantial bridge that could withstand the spring floods. The pressure was great to keep the railroad operating to get the ore out for the weekly steamer, and I recall that one time when the crews were working around the clock to rebuild the bridge the foreman turned in the time sheets showing 25 hours per day for the workers. The superintendent called the foreman in and said that he could understand 24 hours a day as he knew that the men worked day and night, but how could the foreman account for 25 hours per day? The foreman said that the men also worked during lunch hour and the time sheets were approved.

Cordova was a very small town, at its maximum about 3,000 people, but it was a great place to be raised. There weren't enough in any one group to do much, so we all did everything together, regardless of race, color, or religion. There were two Jewish families, two Irish families, one black family, and one or more of most others—Greek, Italian, Yugoslavian, Russian, natives, and so forth. The Episcopal Church had the library, the Presbyterian Church had a gymnasium, the Catholic Church had a social hall, and the Masonic Temple had a good dance floor. It was great growing up without any racial prejudices but of course I never realized it until years later when I went to the lower 48 states, especially the Deep South and the East, and was appalled at the prejudices against different races.

Our house was one of the larger ones in town, so we often had parties there and the piano teacher would come to play. Mother was often ironing in the kitchen, and one evening my girlfriend and I were in the kitchen talking with her. My girlfriend asked my mother the meaning of the red letter on the shirt she was ironing. Mother said that she sewed the initial of each child onto their clothes to help her in sorting them; the shirt she was then ironing was Edward's, with an *E* on the shirttail, and all of Pat's shirts had a *P* on them. My girlfriend had a good laugh but I was embarrassed.

A movie theater opened in Cordova in 1913 and my oldest brother, Harry, was engaged as the doorman. The job was passed down in our family, and from the first opening until my younger brother Francis left town in 1938 there was always an O'Neill boy as the doorman for the nightly movie. We got to see all the movies, although they were all silent movies during my stint as doorman. However, the organist was excellent, and I always enjoyed the music as well as the films.

Mother had one of the early box cameras put out by Eastman Kodak, which she used to record important events; she also hired a professional photographer occasionally. When I was about 13 she had a photographer take individual photos of each of us, which were assembled into one print over two feet long with Mother and Dad and all 12 of us lined up in order. It is a great picture and one I have always enjoyed, and it always occupies a readily visible place in our home. When I was two or three, someone took a photo of me on the dock and had it blown up to 14 by 22 inches. Mother gave copies of most of the photos to Dr. Council, who delivered most of us, and when he died in Juneau his daughter sent the pictures he had to Cordova where they were put on display in an O'Neill section in the small museum there. When our son Kevin was seven, we went to Cordova on a visit, and while we were gassing up a borrowed pickup truck close by, he and our daughter Erin visited the museum. Kevin then came running out, telling us very excitedly that there was a large picture of him in the museum. It was a copy of the big one of me, and it was amazing how much we resembled each other.

I mentioned earlier about the Sunday dinners at home with guests. I was too young to fully appreciate how many of the interesting people who came through Cordova visited our home. I do remember very well Father Bernard Hubbard, the well-known glacier priest who explored many areas

Entire family from left to right: Dad, Mother, Harry, William, Alice, Philip, Edward, Michael, Patrick, Florence, Francis, Margaret, Teresa, Rose Mary.

in Alaska, and I also well remember serving Mass for him. The delegate to Congress Tony Dimond was a guest whenever he was in town, and his talks about politics and life in Washington, D.C., were always interesting. Many of the early-day pioneering aviators came to dinner; these included Carl Ben Eielson, who started the first air service in the interior, and later his father, who came up from Minnesota when his son crashed in Siberia. I remember thinking on several occasions that it would have been nice if the guests had opened the floor to questions after they spoke. However, children were to be seen and not heard and could not speak unless spoken to, so questions went unasked, although I would try to slip in a question while guests were departing. Our local priest, Father Mac, would call around on Sunday afternoons to ask the ladies what they were having for dinner, and then call back the one who was having something he liked and invite himself to dinner.

Mother often said that fish and guests smell after three days and that once we left home we would be considered guests if we came back and could never live at home again. She said that if we were destitute, we could pitch a tent in the yard and come inside to eat, but not live in the house. Years later when I went back to Alaska to visit my parents in Anchorage, which I did almost every year until Mother died, I would occasionally stay with her and Dad but never stayed longer than two days. Mother would get unhappy that I would not stay longer, but I would remind her of what she had told us about guests, and she wouldn't complain about my leaving.

Occasionally during the summer day trips organized on a Sunday to go by small boat out to Strawberry Point to pick wild strawberries or to a fish cannery a few miles out of town where we could picnic. One Sunday at the cannery, which was upriver a few miles from the coast, they served watermelon, a rare treat which we all ate in large quantities. On the way back we were late for the tide, and the boat was stuck on a bar for several hours. Everyone had to go to the bathroom, but there were no facilities on the boat and we were crowded onto the deck. I wonder when I think about the incident now why some of the adults didn't organize with a blanket or something so that people could take turns relieving themselves, but they didn't. When we finally pulled up to the dock, people jumped off the boat and took off running. One short, very heavy woman ran behind the closest car and squatted; a local butcher then jumped into the car and drove off, leaving the lady doing her business. She was so furious that she never went into his butcher shop again.

Cordova had very frequent rain in the summer and heavy snowfall in the winter. It was often said that three continuous days of sunshine at any time was a great summer. Occasionally the snow would accumulate to five feet or more in depth, but usually there was a foot or more of snow followed by rain and then freezing, so it was not uncommon to have icy streets. Lake Eyak was a little over a mile from town and extended for seven miles, so it was frequently a wonderful place to skate or be pulled on skis behind a car, after it froze over but before a rain that would leave the surface too rough for skating. When the streets were icy, we would put on our skates at home and skate downtown or to the lake on the icy roads. Dad liked skating and would go out alone on Sunday when the ice was good and skate the length of the lake. One time he broke through at the far end where it wasn't thick enough. He kept trying to get back out but the ice kept breaking, so he had to keep trying for some time before people heard him and helped him back onto the solid ice. He was badly cut, and his clothes were in tatters from breaking the ice continuously. He was really very lucky to have lived through the ordeal.

All of us except Dad had smallpox when we were little and were quarantined for several weeks, so Dad would come to the house regularly with bags of food that he would leave out in the yard for us. Fortunately most of us were not terribly sick and none of us had many noticeable scars, but it was quite an ordeal for Mother, with all of us underfoot for so long.

One year in school they did not have enough girls for domestic science and had too many boys for manual training, so we all had to draw straws to even out the classes. Mike and I drew short straws, so we had to take domestic science. We learned how to darn socks and to sew on buttons and patches, which was useful to me later on when I was working in isolated mining camps. We also each made an apron for Mother. The cooking classes consisted mostly of making pies, cakes, and cookies. I also learned many of the basics of helping out at home, so I didn't hesitate to hire out as a cook on a small prospect crew several years later in Fairbanks, although I did stop at a bookstore on the way to the job to buy a cookbook.

The dentist I went to in Cordova was an older man who had left his business and gone to seek his fortune in the Klondike gold rush of 1897. He didn't do well mining and later on went back to dentistry with minimum equipment and, apparently, rather poor eyesight, as evidenced by the pieces of cotton under old fillings found by later dentists. He did not have X-ray equipment and would occasionally pull a good tooth when looking for a bad one when I had a toothache; I have had to live with partial plates for most of my life. A dentist friend in Fairbanks did some work I needed when I was in college. I was broke as usual and could not pay until the end of the next mining season, but the dentist liked his whiskey and whenever I had enough money to buy a bottle he would work on me in the evening. We would have a drink or two before, during, and after the work and then go home happy. The crowns he put on lasted for many years.

Violent storms were not unusual in Cordova, and a blizzard developed the day our basketball team was to leave for Anchorage. The ship we were to go on was due to depart at about eight in the evening. The snow and wind had increased so much by midday that the school was closed. The coach gave the student who lived the farthest from the ship a long rope so that he could pick up the other members of the team to tie on as he made his way to the ship. The one who lived the farthest was fortunately the biggest and strongest of the team, so he broke trail through snowdrifts three or four feet high, and we all followed along, holding on to the rope. We finally arrived at the ship, which was rocking back and forth in the howling wind. The longshoremen were having a tough time unloading the cargo so the sailing was delayed, and the coach told us to go down below and go to bed. I dozed off but soon awakened feeling very sick, so I put on some clothes and went up on deck and vomited over the side. A man came along and asked if he could help me. I thanked him and said that I thought I would be all right but that if the sea got any rougher I didn't know if I could stand it. He replied that the sea would probably get much rougher as we were still tied up to the dock. I felt like a fool and I guess I was. We arrived at Seward the next morning and took the train to Anchorage. While we were in Anchorage the team members were each invited out to private homes for dinner. I was assigned to a very big, nice home. When I sat down to eat, there were more forks, knives, and spoons than I had ever seen at a place setting. I did not know where to start so decided I had best wait to see what others did. It was a great experience and one I was fortunate to have had. The storm kept up and the railroad was closed down, so we missed the return boat. It would have been another week to the next boat, so our parents got together and chartered the pilot Harold Gillam to come and get us. The weather cleared the day we flew to Cordova and it was a beautiful trip. It was my first airplane ride and the start of a long association with Harold Gillam and aviation.

When I finished high school at age 16 I told Dad that I would like to go to college. He said he hoped that I could make it. I knew he meant financially but often wondered if he didn't mean intellectually as well. It was during the Depression of the early 1930s; the copper mines had shut down as well as the railroad, and business was bad and getting worse, so he certainly could not help me financially at the time. Also, he had helped two of the older ones go to the lower 48 to college, but they had a good time and never finished, so he vowed that he would not help any others. He did suggest that I write to Judge Bunnell, an old friend who had recently opened a small college in Fairbanks, and that perhaps he could help me. The judge wrote back that he could give me a job as a janitor in exchange for my room and board, and that if I could make enough in the mines during the summer season for my tuition and other expenses I could get an education. I wrote thanking the judge and told him that I would be there in the fall. The day after high school graduation on May 7 I went back to the mine where I had worked the previous season. The pay was good for those Depression times, five dollars a day for ten hours a day, seven days a week, with a bunk in a large tent with five other men and three good meals a day. We were many miles away from any village and the only money you could spend was on gloves and an occasional roll of Copenhagen snuff, so by mid-September I had $600. En route to Fairbanks I stopped by Cordova; Dad told me that things were very bad at the store so I offered my $600, which he said would help, and I cabled the college that I would have to postpone for a year. There weren't any jobs to be had and Dad had more help than he needed at the store, so I scrounged around finding odd jobs. People all had coal stoves

Zenith plane, Copper Center, 1933. Harold Gillam (right) and Patrick (second from right).

and many of them were in second-floor apartments, so I often got jobs carrying coal in a washtub up to their coal bins for two dollars a half ton. I also drove a taxi on the one day a week that a ship was in port. One day when I didn't have anything to do I walked out to the lake where Harold Gillam had a hangar. He was working on a plane, so I started helping him and by the end of the day he hired me at the grand sum of $30 per month. Harold had three planes with one other pilot and was pioneering an air route up the Copper River Valley to provide service after the railroad had shut down. I would meet the planes as they arrived in the late afternoon, drain the oil, cover up the engines, and tie the planes down. Then at six in the morning I would go out to the hangar, start a fire in a little shack to heat up the oil, and then put a fire pot under the covered engine. When Gillam showed up with passengers and cargo, I would load up the plane, put the warm oil in, and uncover the engine and help get it started. Two of the planes had inertia starters, so I would climb up on a strut and crank the motor until the engine started, then jump down and kick the skis loose or move the tail back and forth to get the plane moving. For one of the planes, a Swallow biplane, you had to spin the propeller to get the engine started. It took me awhile to do that well as I was afraid of getting hit by the propeller as I swung it while standing on ice. Gillam would take me along on some flights when he had cargo to drop or after heavy snowfall when he would land on small lakes or short runways. I would compact the runway snow by trudging back and forth on snowshoes while he was off visiting friends or seeing the schoolteacher. Often we would drop cargo to people in isolated areas. We would take off the door of the plane and I would throw out the bundles while Gillam flew as low and slow as he could. Usually I could get the bundles fairly close to the houses, but we had one customer who complained that I was dropping the bundles too far from his house. The next order from him was a frozen quarter of beef, so I let it go just as the plane was approaching the house and the beef went right in the front door, breaking it and damaging many things. We had serious complaints and very few orders after that from that customer.

Dad often accepted paintings or other valuable items when people could not pay their bills, and at one time he had several paintings by Sydney Laurence, who later became a well-known artist. He had them stored in the basement, and they were all destroyed in a flood, a terrible loss with no insurance.

Gillam was a fearless flyer and I often thought he had ice water instead of warm blood in his veins. He would fly through the worst possible weather as calmly as if he were out for a nice Sunday drive, and with very few instruments as none were available at that time. On more than one occasion when I was with him in terrible weather with no visibility, flying where I knew we were in a valley with mountains on both sides, I would be on my knees praying, but somehow he always got through. After a month with Gillam he let me use the car, so I would go out to warm up the engine, put the warm oil in, get the passengers and cargo ready, and then go in to town to pick him up. He would either be in the Model Café having breakfast or across the street in his apartment. One morning he wasn't in either place, so I inquired around as to who he had been with the night before. I then went to the lady's apartment, which was on the ground floor. No one answered my knock but the window was slightly open, so I raised it and the blind and told Harold that the plane was loaded so let's go. He cursed me and fired me and said to leave the car with the key in it, which I did. Three days later he called and hired me back, saying that he couldn't find anyone else who would slave like I did for a dollar a day. That was the only time I was ever fired, and it taught me to mind my manners as far as the boss's personal life was concerned. Gillam had acquired a Siberian sled dog pup, and it flew with him for awhile but then got too big to hold on his lap, and as she got scared in the cabin alone in bad weather he gave her to me. Sertza was her name, and she was a great dog and wonderful companion.

Gillam's well-earned reputation as one of the best of Alaska's bush pilots continued to grow. He had a few crashes but never hurt anyone until the end of his career. The natives called him "chill'em, thrill'em, spill'em, but no kill'em Gillam." He had a mail route one year from Fairbanks to the Kuskokwim region and had an enviable record for completing his flights. In very bad weather other pilots would be sitting on the ground, but Gillam would arrive and leave while the others were waiting for the weather to improve. One time when he was overnighting at McGrath, the winter weather was so bad that everyone thought it impossible to even think about taking off, but Gillam got into his plane, which was on skis, taxied over to the river, and then taxied down the river a few miles until the ceiling lifted enough for him to take off. His luck ran out, however, during World War II. He had engine failure and had to crash-land in southeast Alaska. After waiting several days for search planes that didn't appear, Gillam left what little food he had with his passengers and walked out to the coast, taking his parachute for some protection. Boats passed without seeing him, and he finally took off his long red underwear to make a flag. This was spotted three days later, but Gillam had frozen to death in the meantime. However, he had left a diary with instructions on how to get to the plane, enabling a rescue group to find the crash site. One of the passengers had died from injuries sustained in the crash, but the others, although very thin, were rescued.

My brother Philip worked on Dan Creek in the McCarthy district, near where Bill and I worked on Chititu Creek. The winter I worked for Gillam, Phil had sent word at the end of the mining season that he was going to spend the winter trapping and would be out of touch until March. By the end of March Mother was more worried each day and asked me to tell Gillam that if she did not hear from Phil soon she wanted to go to McCarthy with him. The next day when I had the plane loaded for Gillam's regular flight to McCarthy there was room for me, so I went with him. After we landed, I was on the wing putting gas in the tank when I saw a dog team approaching and saw that it was Phil. I asked him to write a note to Mother for me to take back, as he had some things to take care of before he could leave. Phil had a left-handed, upside-down way of writing that was easily recognizable, and when I handed the note to Mother she was unbelievably happy and relieved. While we were talking afterward about the importance of letter writing, Mother asked me to promise that I would always write her regularly, especially as she had a strong feeling that I would be the traveler of the family, and in addition to keeping her informed of my doings she wanted me to tell her about places and sights that she had never seen and probably never would. I did turn out to be the most widely traveled of the family, and I did write to her regularly and sent pictures and told her about places I had visited. I always wrote on Sundays and missed very few in the 29 years I was away from home before she died. Actually during the war I frequently wrote more often. At most of the airfields the Gray Ladies had a place near the operations office where the pilots had to check in and out. The ladies would serve sandwiches and coffee, and it saved a lot of time for pilots passing through. Many of the ladies had a desk set up with post-cards and stationery, and they always insisted that you write a note to your mother before they would feed you. At first I would argue a bit, saying that I really did write my parents every week, but they always replied, "That is what all the boys say, and if you really want coffee and a sandwich sit down and write a note to your mother and we will mail it for you." So Mother was always happy when I was on frequent cross-country flights. At Amarillo, where I was based for over a year, one of the Gray Ladies named Mrs. Wells looked a lot like my mother; one time she put a note in with my letter so my mother wrote her and they corresponded until Mother died, and then I took over writing Mrs. Wells until she died. Mother was a great letter writer, with excellent penmanship in a clear, firm style that never changed up to the day she died. Her day started with an early breakfast and letter writing for an hour or more, then into her easy chair in the living room for a cigarette while she listened to the news. I went home for a brief visit almost every year except during the war, and the last time I saw her was in 1959. We were having breakfast when she said that she was praying for a quick and easy death. I was shocked and just could not believe what I was hearing. Mother was only 75; she had some health problems, mostly related to her heart, but overall was reasonably well, as far as I knew. She said, however, that she felt she had accomplished her mission on earth and did not want to outlive Dad, who, although three years younger, was in poor health as the result of a stroke several years before. She had buried three of her children and did not want to bury any more. She went on to say that

Our parents' 50th wedding anniversary, 1953.

Mother and Dad's 50th wedding anniversary, 1953.

she had been active in many community affairs in addition to the Catholic Church but had lost interest in almost everything and was worried about lingering illness, so she had made her peace with God and was ready to go. Arguing with her was of no avail, and I left very sadly a few hours later to return to New York. Two days later I went to Bogotá, Colombia, and the day after my arrival received a telegram saying that Mother had died. She had had her usual breakfast, written several letters, gone into the living room to see the news while she had a cigarette, and had gone to permanent sleep. I cabled that I would return as quickly as possible and to please hold the funeral. I arrived three days later in time to see her before they closed the coffin. She looked so peaceful that I knew we should not be distressed but should just thank God for the many years of love and affection from such a wonderful person. That evening after dinner when all nine of us living children were preparing to leave Dad's house, my sister Teresa began crying uncontrollably. My sister Rusty told her to stop crying or she would have all of us crying. Teresa said that she did not think the rest of us could feel as badly as she did because she had been Mother's favorite. Rusty replied that she could not understand how anyone but she herself could have been Mother's favorite: she had lived close to her all her life and had had 15 children (compared to Mother's 12) and was closer than anyone else to Mother. Rose Mary, the youngest, said that everyone knew that the baby of the family was always the favorite. Philip said that the black sheep of the family was always the favorite, and so on it went around the room, with each one saying why they thought they had been Mother's favorite. All the time I could not believe how any of them could think that they had been her favorite when I knew beyond doubt that I was. I had taken classes in school in subjects that she wanted to learn, like Latin, and we had studied together. I had been by far the most widely traveled and the most faithful letter writer, telling her about places I had been and things I had seen, and had sent pictures that helped her satisfy her desire to know about places she wanted to see but had not been able to. We carried on a wonderful correspondence; I found out later that one of the last letters she wrote on the day she died was to me. Whenever I think about it I am always amazed how every one of her children thought they were her favorite. She really was a wonderful person.

When Grandpa O'Neill brought my parents to Cordova in about 1908, he obtained a small shack for them and their two babies close to the Red Dragon Library, which had been built by the Episcopal Church. The church, as I have mentioned elsewhere, had the only library and reading room in town, and Grandpa was one of many who gathered there in the evenings. He became a good friend of the Episcopal minister, Eustace P. Ziegler. After my parents arrived, Reverend Ziegler used to visit with them, especially when Grandpa was in town from the construction camps. Reverend Ziegler often preached to my parents and Grandpa and tried to convert them from Catholicism; he had no success but he kept on trying. Grandpa was a foreman on the construction of the railroad known as the Copper River and Northwestern Railroad from Cordova to Kennicott, where the copper mines were located. The construction camps moved along as the construction progressed, and Grandpa would often take Reverend Ziegler, who was an amateur artist, out to the construction camps for a few days at a time, where Ziegler would paint the glaciers, the mountains, the packhorses, the Indians and their cabins and caches, and anything else he saw of interest. After several years he left the ministry and went east to study painting and later returned to Alaska and became a famous Alaskan artist. Ziegler kept in touch with my parents after Grandpa was killed during the construction of the railroad and later became a Catholic himself. I heard later on that Ziegler's wife had inherited a very substantial sum from one of her relatives and that they had moved to Seattle, so I assumed he had retired. However, several years later when I was working in New York I was visiting Seattle on one of my frequent trips to Alaska and was told that Ziegler was still painting. I had always wanted one of his paintings, so I called and went to visit him. He was painting in a loft looking out over Puget Sound on the Seattle waterfront. He put on a pot of coffee, and we had a great time reminiscing about Cordova, my grandfather, my parents, and others in my family that he remembered. He was in his late eighties but still spry, and I said that I was surprised to see him painting, as I had thought that he would be retired. He said that his wife had inherited a lot of money and wanted him to retire but that he was damned if he would be a gigolo and live off her money, so he was going to keep on painting and intended to do so as long as he was able. He was working on a painting about 16 inches by 30 inches and said that it was one he was doing on commission for $15,000. Then he showed me a smaller one he had just completed that he had sold for $10,000. We talked on some more, and then I said that I had really enjoyed visiting with him but that I had to leave. He asked if I had had anything in particular in mind when I came to visit him. I replied that I had always remembered him and what a great friend he had been to my grandfather and my parents and that I really enjoyed hearing about the early days and hearing him tell of his memories of them. He responded that he had rather thought that maybe I wanted one of his paintings. I told him that in all honesty I had always admired his paintings and dreamed of the day that I might be able to afford one but that day had not yet arrived in view of what he was getting for his paintings. He asked if I had seen any of his paintings that I particularly liked, and I told him that I had seen one of packhorses crossing a river with Mount McKinley in the background. He told me a few more stories of his visits to the construction camps, which I really appreciated as Grandpa had been killed before I was born and I always liked to hear about him, and then I left. About three or four months later I received a crate from Ziegler with a 32-inch by 36-inch painting of exactly what I had mentioned—the pack horses and Mount McKinley. Enclosed was a note saying that he had greatly enjoyed my visit and the wonderful memories it had brought back, as his time in Cordova and his friendships there had been the most important and enjoyable of his life. He also included a bill for $500. I was so surprised and pleased, and I have had that painting for over 35 years now, and not a day goes by that I don't enjoy seeing it.

It was great growing up with six brothers and five sisters. Harry was the oldest, born in January 1906, and Rose Mary was the youngest, born in September 1922. We averaged 17 months between us with the longest space being 22 months and the shortest 14 months. It was amazing how many different personalities and characters there were among us. In order we were Harry, William, Alice, Philip, Edward, Mike, Patrick, Florence (Rusty), Francis, Margaret, Teresa, and Rose Mary. Being so close in age we all played and worked well together and got along very well, yet still were quite independent in our various activities. Harry and Alice went outside to college for awhile and then came back and worked in the store with Dad. Bill, Francis, and I worked our way through the University of Alaska, which until 1935 was known as the Alaska Agricultural College and School of

Mines. All three of us graduated in mining: Bill in mining geology in 1934, I with two degrees in mining engineering in 1941, and Francis in mining engineering in 1942. Five of us were in military service during World War II, and our parents proudly had a banner with five stars on it in the window of their house. Bill and Phil were in the Navy, Bill as an officer in the South Pacific and Phil as a chief petty officer in the Aleutian Islands. Alice was in the Women's Army Auxiliary Corps with the Signal Corps in Washington, D.C. Francis and I were both in the Army Air Corps, Francis as an aircraft maintenance officer with the Eighth Air Force in England and I as a pilot at various bases in the States as engineering officer, operations officer, engineering test pilot, and flight instructor. All of us except Alice were married; three of us, Phil, Margaret, and I, had failed first marriages but very happy second marriages. Phil and Margaret did not have children, but I had three sons, Patrick Jr., Timothy, and Frederick, each of whom have had successful careers in computer-related consulting businesses. Sadly, Fred died of cancer at age 50 in 2005.

All 12 of us in my family grew to adulthood, but some died fairly young. Mother was in Fairbanks, staying at the college while getting medical attention for Margaret, when we got word that Edward had frozen to death in Valdez on December 5, 1935, at the age of 23. I had the very sad task of breaking the devastating news to her and then went with her on the train to Seward, where Dad met us. Harry died in a fire less than two months later in Juneau on January 26, 1936, at the age of 30. Alice died of heart failure in Anchorage in 1946 at the age of 38. Bill drowned in Liberia in 1974 at the age of 67. Phil died of cancer in Anchorage in 1985 at the age of 74. Margaret died of heart failure in Seattle in 2004 at the age of 84. Francis died of heart failure in Lemon Grove, California, in January 2005 at the age of 86, and Mike died of heart failure in Anchorage in March 2005 at the age of 91. So at the time of this writing in 2006 there are four of us left, Rusty, Teresa, Rose Mary, and me.

Florence was named after our mother, but when she was born Dad observed that her hair was rust-colored and she was known as Rusty from then on. She was the only one with red hair in the family. Rusty was also a very good student; one time when we went home with our report cards Rusty was crying because she had only gotten a B+ in one subject instead of her usual A. As Mike and I would dance with joy whenever we got a B we could not understand the crying. Rusty also outproduced all of us, as she had 15 children and at the present writing in 2006 has 48 grandchildren, 64 great-grandchildren, and 5 great-great-grandchildren, a total of 132 direct descendants. The house where Rusty and her husband raised their 15 children in Anchorage, Alaska, was not very big and had only three bedrooms, so most of the gang slept in bunks in the basement. They had only one bathroom, so they divided it off into three areas so that one could be in the shower while another was at the sink and a third at the toilet at the same time. The combination kitchen and dining area had the large table from our home in Cordova but could only be opened enough to seat 15 at a time. The kids often had friends over, and every Sunday any of the relatives who went to 7:00 a.m. Mass were invited for breakfast. Bill would be cooking pancakes while Rusty cooked eggs and bacon and the kids served juice and coffee, and one could eat whenever there was space at the table. Whenever I was in Anchorage on a weekend I went to their house, and it was always a fun time and a good incentive to go to early Mass. I used to say, before Anchorage got so big, that on my infrequent visits when I walked down the street I would often pass kids who said, "Hi, Uncle Pat," without my knowing who they were.

Rusty and Bill rent a big hall in a local hotel for their annual Christmas party, and they draw names so that each member of the family gets a present. My wife Sandra and I went to Anchorage for their Christmas party one year when they had almost one hundred descendants. It was really overwhelming to be in a room with so many blood relatives; in addition to Rusty's family some of the other brothers and sisters were there with their children, so it was quite a mob. All of her 15 children live in the Anchorage area except for one daughter and her family who live in Montana. Nine of Rusty's children graduated from college, all on scholarships, one as a dentist and the youngest with a Ph.D. in mechanical engineering, and all are alive and doing well at the time of this writing in 2006. Rusty and her husband filed on a homestead in the Matanuska Valley out of Anchorage, and the family planted a large area every summer. When it came time for harvest, the entire family worked and the earnings would be designated for one child or another's college money.

My brother Mike was the only one of the family to stay with Dad in the business, and after the bank foreclosed during the Depression of the 1930s

Sandra and Patrick on the Iditarod Trail at Knik, Alaska, 1993.

and took over the store in Cordova, Mike went with Dad to Anchorage where they got into the liquor business—both bars and stores. Mike retained an interest in the business, although he retired several years before he died. Mike was a natural for the retail business. He always had a ready smile and a cheerful greeting for everyone and a great memory for names, besides always being upbeat and optimistic. When he was 80, Mike was a pallbearer at an old friend's funeral. It was the old system where the pallbearers lowered the casket into the ground with straps. There was snow and ice on the ground, and when the casket was almost down Mike slipped and fell onto the casket, breaking fingers and ribs, but he was laughing along with the others as they pulled him out. Not long after that incident Mike had heart troubles and the doctors decided on a quadruple bypass, but Mike was a thin, small man and they could only find enough good veins in his legs to do a triple bypass. Mike was soon out walking and greeting everyone with a big smile and his usual cheery "Hello!" as he had always done. A year or so later he had bladder cancer, so they took out his bladder and fitted him with a bag, and Mike told all his older friends that this was the way to go: he didn't have to get up frequently during the night anymore to go to the toilet like the rest of us old-timers, and the only thing he had to worry about was the bag freezing up if he stayed out too long in the very cold weather.

My first job in mining was working under my brother Bill. He was a hard taskmaster, but I learned a lot working with him. He was still in college my first year there and was very helpful, and we collaborated on some consulting projects in later years. I was 18 and Bill was eight years older, and he would get so exasperated when we entered a saloon and he was asked for identification but I wasn't. Many years later after funeral services for Dad, five of us brothers, Bill, Phil, Mike, Francis, and I, were standing at the end of the bar at one of the family bars when a man came over to talk to us. He said that he and his friends at the other end of the bar were trying to figure out how the five of us stood in age and asked if we would tell them the truth, which we agreed to do. They said that they did not know the one from New York, which was me, but that obviously he was the oldest; Bill was so pleased to be considered younger than me that he bought us all a drink. Bill was a consultant for many years and would often visit or stay when he was passing through New York on his way to South America or Africa on different jobs. Phil was the comedian and a great storyteller and always very nice with everyone. I well remember when the Charleston was in vogue and Phil with his bell-bottom slacks was putting on a great show with one of the cuter flappers of the day. Some of the nicest letters I ever received about promotions or new jobs were from Phil. He would visit Washington, D.C., fairly often on business for the Matanuska Valley Electric Association and always came to see us in Connecticut, which was great fun. He often promised Sandra's mother, Hiawatha, that he was going to clean up one of his Indian friends and bring him out to her as he did not want her to be lonely. When we were younger, Phil and I would go to visit our grandmother O'Neill in her apartment in Seattle, and the first thing Phil always did was to search her closets and under the bed looking for a man, as he was sure that she had one around someplace. She liked being teased and greatly enjoyed Phil. Francis, after the war, went to work for Dad in accounting, which he enjoyed more than mining; he later moved his family to California to put them in Catholic schools and worked as an accountant for a food chain and did very well. Teresa went to the University of Alaska for a year and then met and married Norm Hartung, who was in the Navy in underwater demolition during the war and later worked in Palmer, Alaska, until he died. Rose Mary married a young man in the service during the war and moved to Oregon, where she has lived the rest of her life raising five children who have been very good to her in her later years. Margaret married an Army man after the war, and they were stationed in various places and then retired and lived in Seattle the rest of their lives.

Bob Korn was a big man who worked for many years in the post office in Cordova. He was always very jovial, very pleasant, and well liked by everyone in town. I left Cordova in 1933 and never saw Bob again until 1965, when I was walking down Seventh Avenue in New York City one evening after work to get some hooks to hang pictures in my apartment on West 55th Street. I could hardly believe my eyes when I saw Bob Korn standing in front of a small hotel. I said, "Gee, what a pleasant surprise to see you, Bob; what brings you to New York?" Bob replied, "I haven't been to New York for over 30 years so decided it was time to come for a visit. I checked into this hotel an hour ago, cleaned up, and came out to look for someone I knew to have dinner with and along you came." I said that I could not believe that he expected to find someone he knew in a place as large as New York City, but he said that he always had good luck running into friends wherever he went, so, "Let's go eat," which we did. After a very good meal we went to two different nightclubs and had a really good time together. I am still amazed at how unusual it was to run into him and how nonchalant he was about seeing me on a very busy street in New York.

I was visiting Dad in Anchorage in 1964 and left a few hours before the big earthquake on Good Friday. The D & D building, which had the principal bar of the family business, collapsed, killing two people, as that entire side of the street dropped by several feet. The area by Cook Inlet where our sister Rusty lived with her family had some large crevasses open, and two children who lived in the area fell into one of them and were lost. Dad had had a stroke several years before and could not talk, but the housekeeper told us that he fell in the kitchen during the quake. She had just finished waxing the floor and Dad was sliding back and forth, laughing and having a great time during the several minutes that the house was rocking. Phil was running down a street when a scantily clad lady ran out of a building and started running next to him. Phil assumed that because of the earthquake she had fled her house not realizing that she had so few clothes on and asked her if she hadn't forgotten something; she ran back into the building and came out, still scantily clad, but with her purse. There was tremendous damage in Anchorage, Seward, and many other towns even as far away as

Family-owned D & D Building after the 1964 earthquake.

Cordova, where the entire area rose several feet, leaving much of the small boat harbor with many boats high and dry. They had to build a dock out to deep water and get the boats hauled out. Dad died on August 4, 1964, not long after the big earthquake.

In June 1967 I had the great good fortune of meeting the woman of my dreams, Sandra Elaine Dorris, on a flight from Mexico City to New York City. She was working for Eastern Airlines, and I was on one of my regular trips to Mexico. She would not give me her telephone number, but I gave her mine, and she called me at my office in New York a few days later. From then on we had lunch and dinner together every day that we were both in New York. We did our romancing on the dance floors at the Rainbow Room in Rockefeller Center or at the elegant nightclub at the St. Regis or other nightclubs. We both loved to dance, and we had so many common interests that before we realized it we were hopelessly in love. It was a wonderful romance, and we were married on December 5, 1967, over 39 years ago at the time of this writing. I had arrived from Colombia a day or so before we left for San Francisco for our marriage in Hillsborough; we then went to Acapulco for a short honeymoon, but when we arrived there I became quite ill and could not swallow. A doctor thought I had flu and treated me with aspirin. We went on to Mexico City for Fresnillo board meetings where we joined Mason Smith, with whom I usually went for

Sandra in her uniform as an Eastern Airlines stewardess, 1965.

board meetings. I still could not eat so Sandra and Mason had chateaubriand for two and baked Alaska for two and so on, while I could only watch them. By the time we returned to New York, I had lost 30 pounds and we were concerned that it might be a very short marriage, but I was treated for a viral infection and was soon feeling well, and we have been very grateful for many healthy years together.

We have had a wonderful, interesting, affectionate, loving, happy time with great mutual respect and concern for each other. We both love to travel, and we have certainly done a lot of it. We always seem to have been on the go. For 30 years I traveled about 60 percent of my time, mostly on business. Sandra traveled with me most of the time after we married and was a great help in the mining camps in South America and Mexico, which we visited very frequently. We also traveled to Turkey, Geneva, the Dominican Republic, San Francisco, Vancouver, Toronto, and other places where

I had business. She speaks better Spanish than I do, and the ladies in the isolated mining camps loved having her visit. We had two beautiful, intelligent children, Erin Dorris in 1969 and Kevin Reddy in 1972, both born in Stamford, Connecticut. Sandra's widowed mother, Hiawatha Reddy Dorris, lived with us for over 20 years and looked after the children during many of our trips but also traveled with us quite often. We were blessed to have had her with us for so many years. Sandra's grandfather often commented on how Sandra was always on the go, wanting to go everywhere, and once said that it would not surprise him if she ended up marrying someone from Alaska, which I am glad she did.

Both our children developed diabetes before the age of three, but we took them with us often, even though it was not advised at that time. Sandra was a great organizer and always had everything we needed in the way of medical supplies as well as food. We were determined that Erin and Kevin should do everything and see as much of the world as possible, so we started traveling regularly when Kevin turned eight. We have been around the world and visited all seven continents as well as over 120 countries. I remember when we went trekking in Nepal we each carried a gallon of water as well as a lot of food from home, as we were worried about the water

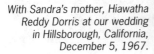

With Sandra's mother, Hiawatha Reddy Dorris at our wedding in Hillsborough, California, December 5, 1967.

and food there. We made at least three trips every year and had wonderful times.

On one trip we went to Japan on business, then to Malaysia where I was on the board of Pacific Tin, then to Australia to look into mining possibilities, and on the way back to Tahiti. My uncle Wilson, who was in the movie industry, had told about filming in Tahiti many years earlier, and he had talked about the beautiful bare-breasted maidens, so I was looking forward to our visit there. However, the Maidenform bra man had gotten there ahead of me, and all the women were wearing bras with their sarongs.

On one of our early trips we went to Wyoming on a wagon-train trip. Each family had a covered wagon with four benches that opened down at night for sleeping. The first night out Kevin and I pitched a small tent to sleep in, but it rained very heavily and we were drowned out, so we spent the remaining nights in the wagon. Since there were five of us and only four benches I obtained extra mattresses and put them between two benches, so Sandra, Kevin, and I had a big area to zip two sleeping bags together. I left a small rock on the floor under Sandra's side, and when she got up I asked how she had slept and she said, "very comfortably," so I said that I thought I had married a princess but apparently not, since she did not feel the rock like the princess in the fairy tale, but she has been better than any princess I could have found. There were six wagons, each pulled by two horses, and the wagons gathered in a circle at night, and we had a big fire and entertainment. One of the wagons was a chuck wagon and two girls prepared excellent meals. The teamster who drove our wagon was a great storyteller and had costumes to go with his stories, as a Confederate soldier, an Indian chief, a frontiersman, a member of the Donner party that was stranded in the mountains during the winter on their way to California, and others. Hiawatha was with us, and she would ride in the wagon swapping stories with our teamster, while Sandra, Erin, and Kevin rode on horseback and I walked, as I was recovering from an operation. There was a family from Sweden on the trip, and they came to visit us in Connecticut later, and then we went to visit them in Sweden a year or so after that. When they were with us, they were impressed with the way I grated raw potatoes and fried them, and they called one day saying that they could not duplicate mine and asked for more instructions. When we were with them in Sweden, the mother took Erin and Kevin out to a beach one day; all the women were bare-breasted and Kevin, who was about eight, was amazed at all the different sizes.

We spent Christmas and New Year's in Austria one year, skiing in Schladming, where we rented an apartment in the same unit as some friends from Colombia. We had not planned to do much for Christmas, but on Christmas Eve Sandra said that we should do something, so we went to the village and saw a nice tree about three feet tall. It was the only good one left, and a lady was about to buy it but decided that five dollars was too much, so we took it. We stopped at a small store and bought ribbon, some decorations, and candles with holders to put on the tree, and Sandra bought a small gift for each of us as well as our friends. It was the only tree we ever had with lighted candles, and it was perhaps one of our most delightful Christmases. We went on to Vienna for the big New Year's Imperial Ball at the Hapsburg Palace. We were very lucky to have been able to get some excellent seats in the main ballroom. The orchestras changed frequently and it was a fantastic evening. We were seated with some people from Japan, and two of the smaller ladies were about Kevin's height (he was 11 at the time), and they never let him rest, taking turns asking him to dance. Both Erin and Kevin are very good dancers, and we all had the pleasure of dancing all evening and well into the night.

Toasting our seventh continent, Antarctica, in 1989. Left to right: Sandra, Kevin, Erin, and Patrick.

Another holiday season we went to Antarctica on a Russian ship. We were surprised to see a blind couple among the passengers, but they had a professional guide with them and it was interesting to hear him describe everything to the couple. When he was describing things to them, we would see things that we had not at first noticed. They went ashore in the Zodiacs along with their guide and everyone else, and he led them everywhere. I recall him describing the penguins and their nests and babies with such great detail that I learned more about them. At the New Year's ball Erin and Kevin danced with the blind couple and were surprised at how well they danced and kept the rhythm. One night we were in pack ice and Sandra spent most of the night on deck with the crew and all her gear, ready to abandon ship, but I went to bed as I thought there wasn't any purpose in both of us being worried. We stopped at Elephant Island, and it was interesting to see the place where part of Shackleton's crew had spent months awaiting rescue.

On one of our many trips we went to Tanque Verde ranch in Tucson, and the family had a great time riding, but as I was again recovering from an operation I hiked in the nearby mountains every day. One day I caught up to a lady in her 50s who told me that she was having trouble trying to keep up with her group. I told her that I would go ahead and when I caught up with the leader I would suggest that he slow down for the straggler. She said to please not do that, because the leader was her mother. When I did catch up with the leader, a very spry lady in her 80s, she noticed the 4000-footer patch from the Appalachians on my pack and said that she also belonged to that club, so we talked about hiking, and I mentioned her daughter having trouble keeping up. She said that her daughter lived in the East and didn't do enough hiking, and so always had trouble when she came to Arizona despite her mother nagging her about exercise. The mother added that she hiked at least five days a week with hiking groups in Arizona and that she was not going to slow down, so the daughter would have to get in shape.

The Explorers Club was founded in 1904 as a multidisciplinary professional society dedicated to the advancement of field research and scientific exploration. I joined the club in 1957 and was fairly active while I was living in New York. I had opportunities to meet famous explorers, such as Vilhjalmur Stefansson, Sir Hubert Wilkins, Peter Freuchen, Admiral Byrd, Bernt Balchen, and many others, and hear about their expeditions and experiences. They had exciting tales to tell and they told them very well. Not long after Sandra and I married the club invited ladies to the big annual dinner for the first time. We arrived at the dinner early, and Sandra was the first woman to enter, and a couple of the older members who did not know about the new policy were near the entrance and challenged us as we came in, thinking that Sandra was trying to crash the party. They were not very happy about women being allowed to the dinner, but it was only a few years later, in 1981, that the club opened membership up to women, and there are now many active women members, one of whom was club president two or three years ago. At the time of this writing there are only a few members who have been in the club longer than I have.

During World War II American and Canadian military people and scientists became concerned with the strategic and economic importance of the Arctic regions. In 1944 a group of senior officials and university people on the Canadian side, as well as scientists and Arctic specialists on the American side, met in New York and agreed to establish a private, binational organization for independent scientific and basic general research on the problems of Arctic living and the Arctic's relationship to the physical, social, and economic problems of the world. Thus was born the Arctic Institute of North America, with headquarters in Montreal and Washington, D.C. I became a member in 1958, I believe it was; was elected to the Board of Governors two or three years later; was chairman of the board for two terms in the early 1970s; and am still a fellow in 2006.

Funding came from both Canadian and American government entities as well as from corporations, foundations, and private contributions. The institute very quickly established itself at the forefront of research on the North American Arctic, and established technical liaisons with Greenland, the Scandinavian countries, and Russia. The journal *Arctic* was launched in 1948 and has appeared quarterly since 1951. A library was established that is still one of the important Arctic libraries in the world. Research stations were established at Kluane Lake in the Yukon and at Devon Island in the Canadian Arctic, and research was carried out all over the Arctic. By the late 1970s, however, government entities as well as corporations were conducting more and more research in house and the institute was unable to maintain its headquarters in Montreal and Washington, D.C. An arrangement was worked out with the University of Calgary to transfer the headquarters and the library there as a Canadian corporation, with a U.S. corporation based in Fairbanks, Alaska. The institute is still active in Arctic research; the strategic and economic importance of the Arctic and its role as a bellwether of global climate change is just as important today as it was 60 years ago when the institute was founded.

In 1972 I was invited to become a councilor of the American Geographical Society and have been active ever since. Established in 1851, it is the oldest professional geographical organization in the United States and is recognized worldwide as a pioneer in geographical research and education. The organization has been noted for pioneering work in exploration and cartography; for amassing the foremost geographical research library in the Western Hemisphere; and for research, mapping, and consulting services for government and business. Geographical knowledge is essential in this age of globalization, and geographers collect, analyze, and communicate information faster and better than ever and provide information of vital importance to decision makers at all levels of business and government. In recent years the organization has also sponsored educational travel around the globe, with informative lectures by leading geographers.

The Ireland–United States Council for Commerce and Industry was founded in 1963 to encourage the development of closer economic, business, and commercial links between the two countries. I joined as director in 1970, I believe it was, and am now an honorary director. The council has been very active and instrumental in encouraging U.S. corporations to invest in Ireland and in lobbying the Irish government to encourage foreign investment with incentives. It has worked very closely with Ireland's Industrial Authority to attract investment in Ireland. In the year 2000 the United States became Ireland's largest single trading partner, replacing the United Kingdom, which had held this distinction for centuries. The council has also been active in bringing students from major universities in Ireland, north and south, to be summer interns in businesses here, and does the same in sending American university students to Irish businesses. For many years the council directors made an annual trip to Ireland to meet government officials and to visit manufacturing plants and other businesses. Sandra and I went on many of the trips and traveled all over Ireland and had great times. The council now also has a very active and growing Ireland Chapter.

During an around-the-world trip when Kevin was 12 and Erin 15 we joined a group at Narita, Japan, for a one-month trip to China, which had just recently opened up to tours. The day we were to leave Narita for Beijing Kevin passed out with a very low blood sugar reaction while he was getting dressed in the morning. We were certain that we had brought glucagon along for such an emergency but could not find it, so Sandra called for someone to get orange juice while I called Dr. Bradley at the Joslin Center in Boston, as well as a doctor in Tokyo who had trained at Joslin. Dr. Bradley told us how to bring Kevin around while Erin ran to the kitchen for more juice. Someone called an ambulance, which we refused to go in, as we were sure that Kevin would respond soon and we knew that we would never make our flight if we went to a hospital. Sandra massaged Kevin's throat to try to get him to swallow the orange juice without choking, which was difficult to do since he was unconscious. However, he finally came around and was taking some food as the doctor from Tokyo arrived with glucagon, so we were able to go with the group. The group consisted mostly of educators and professional people, and we overheard two of them saying what a poor trip this was going to be with kids, especially with one of them being sick. However, it turned out to be a great adventure, and at the end one man stood up in the bus and said that they had all been concerned at the start about traveling with us but that they were thrilled that Kevin and Erin were on the trip as they were so bright and interested in everything. He added that they had brought a different dimension to the journey and made the trip more pleasurable with their enthusiasm. We were the only ones on the trip who did not get sick. Sandra had bought us chopsticks in Japan, which we used at every meal in China, and Sandra took them back to our room to wash after each meal. One time I went back to the dining room for something and saw the waiters wiping off the tables, using the same cloth to wipe the chopsticks, and placing the chopsticks on the table for the next meal, so we think that this is why the others got sick. One evening after a long day we checked into a small hotel in Guilin, and I started to wash my hands at the sink but got my feet wet as it was not connected. The Chinese were so enthused to see children that they always crowded close around to look at Erin and Kevin and try to converse. They were especially interested in Kevin's calculator watch. Kevin said afterward that if that is what happens to celebrities, he did not want to be one. On one flight in China they passed out small blue caps, which everyone saved to wear at Kevin's 12th birthday party in Guilin. I don't remember much about the cake, but I do remember that it is the custom to eat long noodles for good luck and long life on your birthday and that we had lots of long noodles.

One year we went to the Spanish Riviera for the Christmas holiday, and friends from Ireland arranged a rental of an apartment in the building where they had an apartment in Estepona. It was an interesting and much different Christmas that we enjoyed very much. We also took a hydrofoil across the Strait of Gibraltar to Tangier in Morocco. It was a very rough crossing with waves coming almost over the ship, and many of the passengers were sick but we were not. It was very crowded walking on the streets and in the markets and shops of Tangiers; however, the crowds seemed to part for Sandra because she had moved her fanny pack around to the front under her raincoat and looked like she was pregnant, so we did not have the usual jostling. We had lunch in a restaurant where we sat on the floor on cushions and were entertained by a belly dancer. When we left Estepona we visited various towns ending up in Madrid where we stayed for several days. On New Year's Eve we went to a very nice restaurant. One item on the menu was caviar at so much per gram, and Erin asked if she could have some. We said that she could but not too much as it was very expensive, but she kept on asking the waiter for more and more until it was a very large item on the bill. The next day we were buying a leather jacket for me and Erin asked if she could get one but Sandra said, "No, you ate yours last night." A custom in Spain is to eat 12 grapes as the clock is striking midnight, one at each bong. We tried hard but the grapes had seeds and none of us could get to 12, but it was fun trying. Several years later when Erin and Kevin were in college, Sandra and I were on a cruise that stopped in Tangier and we were taken on a tour that stopped among other places at Malcolm Forbes's home. It was like a small museum, with many beautiful things and a tremendous collection of miniature soldiers and armies. The landscaping and the views were outstanding.

One summer, when Kevin was on a sailing and scuba diving trip in the Caribbean, Erin went with us to Alaska and we went to see grizzly bears at Brooks Lodge on Naknek Lake in the Bristol Bay area. There was a two-mile hike on a bear trail to a viewing platform on a river that flowed into the lake, and usually you walked in a group with a ranger carrying a rifle.

One morning, though, we missed the group but were told that we could go alone as long as we made lots of noise, so we banged on a can and hollered all the way. When we arrived at the platform Sandra asked the ranger whether he could hear us coming and the ranger said yes, everyone could hear us, and he was sure that every bear within miles could hear us as well. The bears were mostly in the river just below some falls, and we counted as many as 14 bears at one time. They would catch fish in the river or by diving into the pool below the falls. Quite often some of the bears would stand on the lip of the falls trying to catch fish as they leaped up the falls, and every once in a while one would catch one in its mouth. The bears were of all sizes including some with cubs; the biggest one, who was huge, they called "diver," and he usually came up from a dive with a salmon in his mouth. There were men fishing downstream from the falls, and frequently one of the bears would go towards them and someone would holler "bear" and the men would scatter; even if they had a fish on the line they would cut the line and run. One evening a plane came in and scared the cubs of a female, and they ran in opposite directions, so the mother bear was frantic. We were walking down the beach when all of a sudden the mother grizzly bear came charging out, so we took to the woods and got back to camp as quickly as we could. There was much commotion around the area and everyone stayed inside until the mother found both cubs.

One summer Sandra and I flew to Ketchikan in Southeast Alaska and traveled on Cruise Line West into the fjords and inlets where the larger cruise ships cannot enter. As we neared Petersburg, they announced that six people could sign up for a small boat trip trolling for crabs and fish, and we were the first to sign up. The man with the boat was an interesting old-timer who spoke Spanish; Sandra was engaging him in conversation when he dropped the net, and he left it down longer than usual. He brought up three kinds of crabs, sole, and shrimp, which his wife cooked right away and served with good white wine, so the six of us had a real gourmet feed. When the boat owner was heading for port, we all asked him to make a loop or two before docking so that we could finish the food, which he did. From Petersburg we went on to Sitka and Glacier Bay, then to Juneau, where we met Erin and Kevin on the Alaska ferry. They had boarded in Bellingham and pitched a tent on the deck, where they slept. We went on the ferry with them to Skagway, then to Lake Bennett on the railroad that my grandfather

O'Neill had helped build at the turn of the century. From Skagway we drove to Dyea where the Chilkoot Trail started, and we walked up some to see the trail where Grandpa had hiked on his way to Dawson in the Yukon in 1897. From Dyea we drove to Whitehorse and then on up the Klondike Highway to the Dempster Highway, where we drove to Inuvik. We then went by small plane to Tuktoyaktuk on the Beaufort Sea, the northernmost settlement in the Northwest Territories of Canada. It was quite a trip and very interesting to see the native villages and how the people there lived. All of the facilities in Inuvik—water, electricity, sewer, and so forth—were kept in above-ground insulated housing connecting all the buildings in the village. The village had a large, quite modern hotel, but the only room available was the master suite, which was also used for conferences, so the four of us had a great place for the night. The Dempster Highway was a dirt road for much of the distance and was very muddy, so we had to stay in the ruts or end up off the road in a ditch as several cars had done. We went on to Dawson for two or three days, then made a beautiful trip across the Top of the World Highway to the Alaskan border, and then north to the village of Eagle on the Yukon River. Our Aunt Iloe had gone there as a teacher many years before, when she was just out of college, and had had some exciting experiences. From there we headed towards Fairbanks with a short stop in Chicken, where I had worked on a prospecting job years earlier. After visits to the university and many of the mining sites where I had worked we drove to Anchorage, with a good look at Mount McKinley on the way, to visit some of our numerous relatives before returning home.

When I turned 57 years of age I received a large box of Heinz 57 varieties on my birthday, which was quite a surprise. I never found out how they knew, but we enjoyed it all.

The first seven years of our marriage we had a small house in New Canaan, but eventually outgrew it, having two children and Sandra's mother with us often, so we built a larger house across town that Sandra designed and supervised. It had four bedrooms, a two-room mother-in-law suite, and a pool and tennis court. It was a wonderful house for our family and for entertaining. We did a great deal of entertaining for business and for many of the clubs to which Sandra belonged, as well as for school and family functions. Sandra always decorated it beautifully for parties and especially at the holidays. We enjoyed it for 30 years, but after Erin and Kevin left and

Sandra's mother died it was too big and too much work for just the two of us, so we moved to a town house.

Sandra's mother was named Hiawatha Reddy. She was born in New Orleans, and her father had a very good friend whose wife was expecting at about the same time as his. His friend said that his wife was going to have a girl and that he was going to name her Pocahontas, so Hiawatha's father said that his wife was having a boy and if his friend named his daughter Pocahontas he would name his boy Hiawatha. His child turned out to be a girl, but he named her Hiawatha anyway. Hiawatha had so many fights about her name when in school that she vowed to change it when she came of age, but she said that by then it was such a conversation piece that she decided to keep it. Both of Sandra's parents painted as a hobby, her father

Our home in New Canaan, Connecticut.

in oils and her mother in watercolors. Sandra's oldest sister, who was killed in an accident at age 20, also painted in oils and did it very well. We have several of her paintings and especially like a portrait she did of Sandra at age three or four. Sandra and our children inherited this artistic talent. Erin painted a mural in high school and designed and painted a card of the *Challenger* accident that the entire class signed and sent to the family of the teacher who was killed. She designs and makes special occasion cards as a sideline. Erin and her husband, Alex Argueta, have two children, Haley and Dylan, and they seem to have inherited the Dorris artistic talent. They both make their own special occasion cards. Both Haley and Dylan took me to school at various times to show pictures and talk about Alaska and South America when they were studying those areas. They are both excellent students. Erin designed and assembled a fantastic, very large album with pictures and highlights of my life for my 90th birthday. Each page is decorated and it is truly a work of art.

Kevin did very well in art class in school but decided that he would rather spend his spare time rock climbing, and he is very good at that. Sandra had always had beautiful flower gardens but then got very sick with Lyme disease and was on antibiotics for over three years, so she decided that if she could not grow flowers she would learn to paint them. She started classes in botanical art in watercolor and is doing beautiful work. At the time I am writing this while in Tucson she is taking classes at the Sonora Desert Museum from two master painters from England and enjoying it very much.

Sandra was eligible to join the Daughters of the American Revolution from both sides of her family, as both her mother's and father's ancestors had come over before the American Revolution. Her paternal grandfather, Samuel McClellan Dorris, born in 1860 in Illinois, was a circuit court judge in Nebraska and traveled around in a horse and buggy holding court throughout the state. He was also the blacksmith in Broken Bow, Nebraska. Sandra's father, Herbert Ellsworth Dorris, was born in 1889 in Broken Bow in a sod house with an Indian midwife in attendance. He was in the cavalry with General Pershing when they were fighting Pancho Villa on the Mexican border; later he served in World War I in France in the same regiment as Harry Truman. During World War II he was a major in the Pentagon in the Signal Corps. After World War I he went to Louisiana and acquired a cotton plantation but went broke after three disastrous seasons.

He did, however, meet Sandra's grandfather, Gerard Miller Reddy, and through him met and married Sandra's mother, Hiawatha Reddy. He continued to work until retirement, except for the war years, with the American Can Company, where he lost what was left of his hearing so that he was deaf all of Sandra's life with him. He died just before I met Sandra, so I did not know him. Sandra's maternal grandfather, Gerard Miller Reddy, was born on a large cotton plantation in Natchez, Mississippi, in 1874. He founded and built the town of Bowie, Louisiana, and had a large mill that processed mainly cypress. Cypress was used in building silos as it is a very hard wood that long outlasts other woods, which are prone to rot in damp climates. The entire town was burned down, however, in a fire that started from sparks from a railroad train in the mill yard. He then bought a lumberyard in Napoleonville, Louisiana, which he operated until he died and which the family continued to operate until a few years ago. Sandra's mother, Hiawatha Dorris, was born in New Orleans on August 13, 1906, and married Herbert Dorris in 1924. Sandra was born in Omaha, Nebraska, but lived as a youngster in Houston, Texas, and later in Baton Rouge, Louisiana. After attending Louisiana State University she joined Eastern Airlines where I was so lucky to meet her.

Sandra and a friend had a Girl Scout troop that Erin belonged to for several years; then with another friend she started a Cub Scout troop, of which Kevin was a member. They had six boys; we kept in touch with them afterwards and have seen all six of them married. I was active in the Boy Scouts and took over from the Cub Scout group. I was chairman of the troop and we were fortunate in having great scoutmasters; nine of our scouts, including Kevin, became Eagle Scouts, the highest rank in scouting. I camped out a lot with them, and Kevin and I camped out once a month for a year to get one of his badges. Before scouting Kevin and I were in an Indian guide group at the YMCA. We were in the Ute tribe, and I was the chief of the Utes. It was a great group, and we did a lot of things including camping out often, canoeing, hiking on parts of the Appalachian Trail, and visiting historical places. One time I was scheduled to go on a backpacking trip on the Appalachian Trail but hurt my back and was unable to go, so Sandra went backpacking and tenting with the group of boys, as Kevin had not yet learned to administer his own insulin injections at that time. She was really determined that diabetes was not going to hold Kevin and Erin back from doing everything that the other kids did.

In the early 1990s a very nice family, Sally and Kim Campbell and their five young boys, moved into the second house down from us on Dunning Road. It is a short hilly road that I walked up and down at least four or five times a day for exercise, and the boys would often come out to talk. They were all very polite and friendly, but the middle boy, Peter, was the most interested in talking with me and we would talk at length about Alaska and about mining and my time in the Army Air Corps. When his mother made cookies, he would often bring a few of them over to our house and I would show him pictures and mineral samples and other memorabilia, and we became good friends. He wrote several articles for school about going to Alaska to visit his friend with his big dog and other stories about his good friend Patrick O'Neill. When he was in the second grade, he took me to school for show and tell and later arranged with his teacher for me to go over and show slides of Alaska. Kim was a soccer coach and taught all his boys and many others, and I would often go to watch them practicing and playing games, usually at Peter's invitation. All of the boys are very good athletes and were active on many teams, and Sally seemed to be constantly on the road driving the boys to their various games. Peter's two older brothers were each captain of the high school soccer team in their senior year, and Peter was as well in his senior year. I was recently invited to Peter's graduation party and to wish him well as he was leaving two days later for West Point to follow in his dad's footsteps. I have confidence that he will be an excellent officer and succeed in anything he does.

One of our many great trips was to Africa. We flew to Cape Town, which is a lovely city with many places to visit such as the cape and the wine areas. We went to Johannesburg on the blue train, which was outstanding in service, food, accommodations, and scenery. Some friends whose son Brad was the same age and as a close friend of Kevin's had invited us to visit them, and they arranged safaris to two great camps, Mala Mala and Harry's Huts, where we thoroughly enjoyed seeing so many different animals and got some great pictures. We celebrated Erin's 13th birthday in a boma, I believe it was called—an enclosed circular area made with seven- or eight-foot-high poles and a zigzag entrance that prevented animals from getting in easily. Another girl of the same age was also celebrating her birthday, and we asked for a birthday cake. The head ranger who usually made the cakes was away, so his assistant, who said that he was not much of a cook but would do his best, made a memorable cake. It was flat and heavy but much appreciated nevertheless. We also visited Krueger National Park to see many more animals, and then went to Maun in Botswana which I write about in the section on flying. We spent one night in a thatched-roof building near the river, where the crocodiles came up close to the buildings. There were all kinds of purple lizards and other bugs in the ceiling, so we slept in a pile with me mostly on top so that the bugs wouldn't drop onto the others. We went to a camp called Savuti and saw many animals; however, the weather had been very dry so the animals were suffering and crowding into the few pools of water. Kevin was so distraught that he asked if I could have a large plane fly in water for the animals. It was nice to know that my young son thought I could fix anything. The other camp we visited in Botswana was on a river, and we went spearfishing one night. We saw many beautiful birds, mostly bee-eaters, grouped together on reeds roosting for the night, and they were not bothered by the spotlight that enabled us to see them from the boat. Erin was the only one to spear a fish that night, much to the chagrin of the men on the trip. Not far south of that camp the river disappeared underground into the Kalahari Desert. We then flew from Johannesburg to Nairobi, where because of recent unrest and a coup attempt we were the only ones getting off the plane. We were greeted by army men with machine guns and had to wait several hours until another plane arrived. An army convoy then escorted us all into the city to our hotels, by which time it was the middle of the night. We were the last to be dropped off at the Hotel Norfolk, which was all locked and chained. After five minutes of us banging on the door, they let us in. I was concerned with what I got my family into, but it all turned out all right and we had a great visit in Nairobi. We flew out to the Maasai area, where the people lived in mud huts and were very colorful in the few clothes they wore. Their main sources of livelihood were cattle and selling crafts, shields, and so forth to tourists. We had good accommodations, although one had to be very vigilant to keep the monkeys out. We had a great driver who took us around in an open-top van to see the animals, and he was just as excited as we were one day when we saw a lion kill a warthog.

In the mining section I mention Ray Dunn, who was shot in Andagoya and then quit and went back to New Zealand, married his high school sweetheart, raised two children, and built a successful mechanical shop business. We kept in touch after he left Andagoya in 1954, and he finally called in 1992 to say that his wife had died and the children had finished college, so he had sold his business and was traveling. At the time he was visiting a niece in Washington, D.C., so we invited him to come and see us in New Canaan. Sandra's mother was then in a care center and Sandra stayed with her much of the time, but she managed to be home to fix us nice dinners and we had a good visit with Ray. However, after a week she was wondering how long he was going to stay, so I asked him and he replied that he had originally planned to stay for a week but the food was so good, especially the desserts, he had decided to stay an extra week. I told him that we greatly enjoyed his company but with Hiawatha in the care center it was just too much work for Sandra, so he left. The following year, though, we planned a cross-country trip with visits to many of the national parks and houseboating on Lake Powell. Sandra and I had a great trip driving from Connecticut across the country to the Oregon coast and then down to San Francisco. We visited many of the national parks, where Sandra had arranged accommodations. Kevin was in Montana and met us at Yellowstone for a fun visit, and we stopped to see my sister Rose Mary and her family in Ontario, Oregon. We had been advised that it would be better if there were three people on the houseboat, so we had called Ray and picked him up in San Francisco upon his arrival from New Zealand. In the end we did not rent a houseboat as we found that it was more convenient to go on day cruises, but we had a great time with Ray, with over a month sightseeing on the way back across the continent. From then on Ray came for a month or longer and we traveled together for several years. One time we went to the Maritimes and drove across Nova Scotia and Newfoundland. Twice we met in Europe: on one occasion we traveled around Europe by train, and on another we took the Orient Express from Istanbul to Venice, where we met Ray and visited many of the islands in the Vaporettas and then took the train to Florence and on to Milan and Bellagio. Once we went to New Zealand and stayed with Ray and drove around both islands and then took a cruise from Auckland through the South Pacific, ending up in Hawaii. Ray was very easy to travel with, always had a smile, and when we told him what time we wanted to depart he was always there 15 minutes early with his bag. He always had a bottle of scotch and I had small bottles of wine, so if we decided not to go to dinner we would have snacks and get a Diet Coke for Sandra and play gin rummy in the room. He never complained about anything, although after a month of Sandra fixing us sandwiches with tomatoes for lunch he very politely said that he really did not like tomatoes. He was planning on coming over for Kevin's wedding in Montana when he had a heart attack and died.

When Erin became diabetic, we were in Norwalk Hospital getting her started on insulin and I went to the cafeteria for sandwiches and coffee for Sandra and me. I was a heavy drinker at that time, and Sandra mentioned that this was the first time since we had met that I had not had several drinks before dinner; she was worried that if I kept on as I was going I might end up an alcoholic. I made up my mind then that if Erin could not have sweet drinks, I would stop the liquor and that I would donate the money I saved for research on diabetes, which I have been doing ever since. After two or three months we decided that we would enjoy parties more if I drank something, so we agreed on a glass of wine or beer. Then about 15 years ago when I developed angina the cardiologist suggested that I have a glass of red wine with dinner, so that allows me two glasses which I usually enjoy. We were not very happy with Erin's treatment, but a friend told me about the Joslin Clinic in Boston, so as soon as Kevin was born we went there. Erin and Sandra were in the clinic for a week and I stayed nearby. We went to classes every day where we learned how to live with diabetes, and they taught us how to take care of Erin. We were very appreciative of all they did for us so I made a donation of what I estimated to be my yearly liquor bill. The director of development apparently thought he had a live one and came to see me in New York the following week and asked me to join the Development Committee, which I did. It wasn't long before I was chairman of a committee to raise funds for a new building, and shortly thereafter I became a board member and remained so for 25 years, the last 15 years as chairman. Kevin developed diabetes at the age of two, so we had double the reason to devote our major efforts to raising funds for diabetes research to find better treatment and to avoid the complications that result from years of having diabetes. I went to Boston for monthly board meetings and often at other times for committee meetings and other matters and was on the phone almost daily during all those years, which were challenging and interesting. I was deeply impressed with the dedication and devotion of the research staff, the doctors, the teaching nurses, and in fact everyone in the Joslin organization. They were all dedicated to providing top-quality care to those with diabetes and to conducting research to ensure a future of hope for everyone who was or might become afflicted with the disease. During the 25 years that I was very active in the organization the budget grew from $22 million to $88 million, and the service increased from 36,000 to over 100,000 patients annually. Another capital campaign raised funds to add 72,000 square feet to the facilities in order to modernize patient care areas, increase the scope of research programs, and bring new technology to the escalating fight against diabetes. Research expenditures tripled during that period ,and many important scientific achievements were made in the care and treatment of diabetes patients. The center was chosen to participate in a national diabetes control and complications trial, which proved conclusively that keeping one's glucose levels within the normal range will do much to prevent the devastating complications of diabetes. In 1986 the Joslin Diabetes Center was named a Diabetes and Endocrinology Center by the National Institutes of Health, one of only a handful of such centers in the United States. In that same year Joslin's research division embarked on a major program in molecular biology to identify the specific genetic defects that cause Type I and Type II diabetes and to determine how genes might be manipulated to treat diabetes. The Affiliated Centers Program, initiated in 1987, included 11 affiliates from New York to Hawaii by 1995 and provided diabetes care for more than 60,000 patients in 1994; it has expanded extensively since then. Many new patient education programs were initiated, including a three-and-a-half-day Diabetes Outpatient Intensive Treatment (DO IT) program; an exercise and weight loss program called Fit and Healthy; and Joslin's Pre-School/Early School Age Program, the only one of its kind in the country. I am happy to say that the Joslin Diabetes Center has continued to expand and develop even more dramatically since my retirement in 1995. At the time of this writing in 2006 the center is conducting another campaign to further expand the physical plant and all the activities of this great organization.

Elliott P. Joslin founded the Joslin Diabetes Center and led it for over 60 years. He was followed by three doctors who were with the organization for

Ken Quickel, president of Joslin Diabetes Center, Boston, presenting a medal to Patrick O'Neill, retiring chairman.

many years, Doctors Root, Marble, and Bradley, each of whom contributed greatly. When Dr. Bradley retired in 1987, we were most fortunate in obtaining the services of Dr. Kenneth Quickel, an outstanding physician and exceptional administrator who guided Joslin through the greatest period of growth and progress in the history of the organization up to that time. He was the most well-organized person I have ever known. For me working with Ken was the best working relationship in my 60 years in business. It was a good feeling of mutual respect that developed into a very comfortable and pleasant relationship. We had excellent communication and prepared well for every board or other meeting so that meetings moved along well with successful outcomes for the benefit of the organization.

Living with diabetes is not an easy life; it requires constant dedicated discipline in daily living. Erin and Kevin got off to a good start with the education at Joslin and the management by Sandra in their early years, and they have both done a great job in managing their diabetes. Erin is now 37 and Kevin is 34 at the time of this writing, and both are getting along very well in responsible and demanding positions in business. The development of the insulin pump (where insulin on demand is provided through a needle in the stomach) has been very important to them, and without it they could not handle the long hours and pressures of work and irregular mealtimes so successfully. They each have two wonderful children and are blessed with great spouses who are partners with them in the management of their diabetes.

A few months after Sandra and I married we took off for Alaska from our home in Connecticut in an Alaskan camper mounted on a three-quarter-ton pickup truck. The roof telescoped and was easily raised with a hydraulic pump so that one could stand erect. It had a propane stove, a refrigerator, a 20-gallon water tank with a small pump for the faucet and sink, and also a portable toilet. The two bench seats could be combined for a king-size bed and it was very comfortable. We camped at Niagara Falls the first night out, and from then on we camped by a lake or river or where we could see mountains. After we reached the Yukon Territory, we picked wild strawberries, raspberries, or blueberries every day and caught grayling or rainbow trout nearly every day for dinner the rest of the trip. We traded off on the driving, and the one who was driving when we stopped for the night got to relax while the other cooked dinner. One morning when the engine didn't sound right I pulled off the road to look at it. Sandra said that there was a stream close by, so I got out the fishing pole and she went fishing. When I was through working on the truck, I went to the stream and she had five nice arctic grayling. I said that that would be enough for dinner and started cleaning them, squatting at the edge of the stream. Sandra said that there was one larger fish in the pond and would I please put on a different fly so that she could catch it. She soon hollered that she had caught it. I looked up just as she swung the pole around, and the fish hit me in the face and fell off her hook back into the stream. Sandra was annoyed to have lost the fish, and she probably thought that I should have grabbed it with my teeth when it hit me in the face, but even if I had my mouth open I doubt that I could have bitten it fast enough to hold it. She could not hook it on subsequent tries, but we had enough for dinner anyway.

On that same trip we stopped at Kluane Lake in the Yukon, where the Arctic Institute of North America had a research organization and support base for research crews on Mount Logan and the glacier at the base of the mountain. They had a Helio courier plane that is designed for short-field and high-altitude work. The pilot, Phil Upton, was very well known for high-altitude landings and had made over 300 landings on glaciers and other unprepared surfaces, including many at high altitude on Mount Logan. I was chairman of the Arctic Institute at the time, and Phil took Sandra and me on a flight to the camp on Kaskawulsh Glacier at the base of Mount Logan. High-altitude physiological studies were being carried out at various elevations from 2,900 meters to 5,300 meters on Mount Logan. We flew up the glacier for miles, and as we approached the camp on the glacier everything was white snow, so Phil gradually let down until he touched the snow. When he stopped, Sandra opened the door and jumped out before Phil had time to get out the snowshoes, and she sank into the snow up to her waist. When she sank so fast, I was afraid that she was dropping into a crevasse and might disappear entirely, but fortunately it was just soft snow so we helped her out onto the snowshoes. Being from the South it was her first experience with such deep snow and one that she will never forget.

On that trip we also flew from Fairbanks up to Point Barrow, which is the northernmost settlement in Alaska. The Arctic Institute had a research program with the navy there, and among other arctic animals they had a large polar bear in a cage inside a building. There was just enough room to walk around the cage, and while we were there one of the men held a push broom over the top of the cage to get the bear to stand up so that we could see how huge it was. The bear stood up on its hind legs and grabbed the broom with its teeth, so the man asked me to jab the bear in the ribs with the handle of a shovel, which I did. The bear then put a paw through the fence and took a swipe at me, so I pulled back quickly and almost hit Sandra in the face with the shovel. She had been standing in the corner behind me, but somehow I stopped just in time, with the end of the shovel only an inch or two from her face.

On one of our many trips to Alaska my sister Rusty's oldest daughter, Beth, and her husband, Dick, invited us to dinner at their log cabin home in the wilderness near Knik. Dick had caught a salmon that day near their place, and there was a stream close to the house where they kept drinks cold. Dick filleted the salmon and grilled it on the barbecue and they had potatoes and vegetables from their garden. It was a great feast topped off with a fresh blueberry and a fresh rhubarb and strawberry pie. After dinner they showed us movies in which they had flown up onto a nearby glacier in their small plane in dress clothes; set up a table with a nice cloth, silverware, crystal, and so forth; and had a gourmet meal sitting on the glacier surrounded by mountains. What a fun evening out!

Kevin met his future bride when he was working in Bozeman, Montana, after he graduated from Lewis & Clark College in Portland, Oregon. She left soon after they met to teach English in the Japan Exchange Teaching (JET) program. Kevin went to Japan after he left his job in Bozeman. There, Terri introduced him to a Buddhist priest and his wife, who took Kevin in for three months, teaching him to speak and write Japanese. Kevin then applied for the JET program and while he was waiting went with us to France where the three of us studied French at the Institut de Français in Villefrance-sur-Mer in a one-month total immersion course. Sandra was

At our son Kevin's wedding in Big Sky, Montana, September 2, 2000. Left to right: Dylan, Erin, Alex, Terri, Haley, Kevin, Sandra, and Patrick.

Friends from Japan at Kevin and Terri's wedding.

ill from what we later found out was Lyme disease and was struggling with severe headaches throughout the entire trip. Kevin went off with young people from the class every evening where they only spoke French, so he did better than we did, advancing to French 4 while we went to debutante 2. Then we traveled for a month, practicing what we had learned.

Soon after we returned home Kevin was accepted in the JET program and went to teach English in Japan. He had a small house in northern Japan, and we went to visit him there. It was a nice little house, but it was hard to get up off the floor mats during the night, although it did have the only electrically heated toilet seat that I have ever seen. Kevin and Terri McBride married two or three years later in Big Sky, Montana, while they were both in graduate school. We hosted the rehearsal party in an old barn near Big Sky: the dress code was blue jeans and boots, and we gave everyone straw cowboy hats. The wedding was in a log church in Big Sky with a great view of the mountains out a large window at the end of the church. Ten of their friends flew over from Japan for the weekend of the wedding. The ladies wore their beautiful silk kimonos to the wedding, and during the reception they performed a traditional Yamagata-ken dance, the Hanagasa Odori, with their music and with thin concave straw hats decorated with red flowers. Eight of the ladies that came over have an annual party celebrating the event and always send a picture of themselves with their cowboy hats on. They did the same to celebrate the births of Jackson and Whitman.

Sandra and I were on a cruise in December 2000, and one morning we sat at a table with others at breakfast. During the conversation, which was about traveling, one man mentioned that he was going to Alaska soon to visit some relatives. I asked where in Alaska he was going, and he said that it was just a very small village that he doubted we had ever heard of. I said, "Well try me, maybe I know it," and he said, "Cordova." I told him that I had been born and raised there and knew the family well that he was going to visit. What a small world!

Several years ago I picked up a rental car on a Friday evening at LaGuardia airport to drive home to New Canaan. As I was signing the rental form, the lady behind the counter announced to the line of people waiting for cars that no one without a confirmed reservation would be able to get a car, as they were all booked. The man directly behind me started cursing and said that he had to get to New Canaan in the worst way. I told him that I was

Patrick's 90th birthday. Left to right: Dylan, Alex, Erin, Patrick, Whitman, Sandra, Terri, Jackson, Kevin, and Haley.

going to New Canaan and would take him; he was very grateful and off we went. When we were well underway, I said that his serious distress at not getting a car seemed to indicate that he had an important obligation or a heavy date in New Canaan. He replied that no, his car had broken down early in the week and he had left it for repairs in New Canaan. He needed to pick it up to drive home to New Jersey, where his high school graduating class was having their 25th reunion the next day, and it was very important for him to be there. He went on to say that there were over 400 in his class, the brightest and best class ever at his school, and that 280 or so were attending the reunion. He further went on to say that he was very proud to have placed 114th in this very bright class. I congratulated him on his high standing in the class and said that I was pleased to have been of some help in getting him to his reunion. We talked of other things and after a lull in the conversation he very audaciously asked how I stood in my graduating class. I told him that I was eighth, and he exclaimed, "Gee, you must have been smart." I demurred as modestly as I could, and then he said that in his experience the top students were usually not very sociable or friendly. I told him that that had not been so in my class and that the top fellow had not only been the brightest but also the nicest, as well as being the biggest and the best athlete, and that he was still one of my best friends and very successful in his career. My companion replied that well, that was strange and contrary to his experience, and then started talking of other things. Later he asked where I had gone to school. When I said, "Alaska," I could almost see the thoughts churning, and then he asked how many had been in my class. I answered, "eight," and he exclaimed that I'd really suckered him in but I said not so, on the contrary, that he had been very audacious and rude in asking me how I stood in my class. We had a good laugh and I wished him well at his reunion, and he could not have been more appreciative of the ride. (The top fellow in our class was Woody Johansen.)

For our 25th wedding anniversary I arranged for us to go on a cruise. Sandra agreed to it but when it came time to go she said that she did not like the water, could not swim, and so forth. We argued long into the night, but she finally agreed to go at 4:00 a.m., an hour before the car was to pick us up, so we threw some clothes into a bag and went. We didn't have the

Sandra and Patrick on a European cruise, June 2005.

right clothes but we had a great trip and have enjoyed many cruises since
then. I believe we are up to 25 cruises now. On our first cruise we were
seated with some people from the Midwest and in our preliminary conversation mentioned that we had recently been in France studying French.
The sommelier asked me what wine I would like, and when I said, "cabernet
sauvignon," the lady at the table said, "My, you really did learn to speak
French, didn't you." On another cruise we were seated with an older couple;
the husband had just retired from a brokerage firm and his friends had sent
him bottles of very good wine, one for every night of the trip. Neither his
wife nor Sandra drank, so he and I had fun drinking up his wines. On one
of our recent trips they played a game called "Who doesn't want to be a
millionaire," and Sandra was chosen as the first contestant. They had questions with a prize for each one, and Sandra answered all the questions and
won all the prizes, including the grand prize of a nice painting. Cruising has
been great fun for us.

2 College

THE Alaska Agricultural College and School of Mines was created by an act of the Territorial Legislature in 1917, but it was not until 1924 that it opened with Charles E. Bunnell, a former federal judge, as president. The first year there were six students with a faculty of seven, including the president, who also taught. When I enrolled in the fall of 1933 there were about 100 students, as I recall. The college became the University of Alaska in 1935, and as of 2006 there are 9,380 students enrolled—5,630 at the Fairbanks campus and 3,750 at satellites around the state.

As mentioned elsewhere in these memoirs, Judge Bunnell gave me a job as janitor for my room and board, and I was able to make enough money in the mines during the summer vacation to take care of tuition and other expenses. Close to 90 percent of the students at that time were self-supporting, so nearly everyone had very limited means. A Saturday night date was

The University of Alaska Fairbanks, where Patrick received his education.

Patrick at the University of Alaska, 1936.

often a four-mile walk on the railroad tracks to Fairbanks, a sandwich at the Model Café, and a walk back. Occasionally if we were flush, we would take the bus back; however, the road was very bad and the bus often got stuck, and we would have to help push it. In the wintertime we would ski into town, have a beer at the Chena Bar, and ski back. I would occasionally get work clerking in a clothing store in Fairbanks or tending bar at the Chena Bar. After two or three weekends of work I could take a date to town to the movies or a dance. One time I had a date with a Fairbanks girl when it was 60 degrees below zero. On my way back the bus broke down about halfway out to the college, and I just about froze walking the rest of the way.

There weren't any good hills close to the college for downhill skiing; however, there were some Scandinavians in college who introduced us to cross-country skiing, so this was a favorite winter sport. We would go out skiing whenever we could, often after dark as the days are very short in that area in the wintertime. In December and January the sun would come up about 10 a.m. and set about 2 p.m. In March and April, when the weather was a bit warmer, we would have ski races over 10- or 15-mile courses, depending on the skills of the different groups. I won the class C race one year so had to race with class B the next. There were two skiers in the group whom I knew I could not keep up with; they drew numbers 1 and 2 and I was number 3. It had snowed the night before the race, and as a result of the new snow they missed a turn because they did not see the flag. I spotted it, however, and won the race because the others had been disqualified. When the awards were given out, I was embarrassed to accept the medal but not so embarrassed as to turn it down.

As a janitor I was responsible for cleaning the main floor offices, laboratories, lavatories, and halls. I would give everything a lick and a promise after classes or in the evening during the week and then do a thorough cleaning on Saturday or Sunday—usually Sunday, because I was athletic manager and often very busy with sports activities on Saturday. Occasionally I would be working on Sunday morning with a bad hangover. Judge Bunnell would often go by, and as soon as he realized I was suffering from a hangover he would engage me in the longest conversation. He would have a twinkle in his eyes as he could readily see that I could hardly keep my eyes open, let alone carry on a reasonable conversation. Many years after graduation some alumni members started a campaign to raise funds to erect a statue of President Bunnell as a founder and first president of the university. I was asked to speak at the dedication and spoke about how President Bunnell had given me a job as a janitor and how grateful I had always been

Dedication of the statue of Charles E. Bunnell at the University of Alaska Fairbanks, 1988. Left to right: Patrick, Earl Beistline, and Glen Franklin.

for the opportunity he had given me to get a college education, as I would not have been able to go to college without his help. The next day there was an article in the *Seattle Times* about the statue, and the headline was "College Janitor Gives Dedication Speech." I treasure that article in my files.

One of the best references I ever had was from President Bunnell, when I was applying for officer training in the Army Air Force in 1941. Later on in 1950, when two other graduates from the university and I were promoted to top jobs in the local Fairbanks Exploration Company, President Bunnell wrote an editorial for the local paper saying what a great day it was for the university when three of its graduates were promoted to top positions in the local mining company that had been so helpful to many students working their way through college. The other two men had been several years ahead of me in college, and Bunnell went on in his article about what an outstanding student Jack had been, president of the student body, and so on, and what a great athlete and student Ted had been, and then he said, "and Pat was the best janitor we ever had." Another priceless item in my scrapbook of life. The Fairbanks Exploration Company really helped many of the mining students at the college, not only by giving them summer work in the mining operations but also in training and educating them for careers in the mining industry. I was one of many students who worked in various departments in the company; I worked every job from laborer in the stripping operations, to helper on prospect drills, to panner on the drills, to all phases of underground mining from mucker to millman to assayer to engineer, and in the main office as draftsman, dredge engineer, exploration engineer, and finally, superintendent. It was a great opportunity to learn all the practical basics of mining while studying in college.

During one of my later years in college I worked as a waiter in the dining hall as it allowed me more time for studying than I had as a janitor. There were eight people at each table, and the waiters would carry out the platters of food to set on the tables. One Saturday night when everyone was dressed for a big dance I leaned over to set a platter of potatoes on the table while holding a platter with eight pieces of meat in my other hand. The girl I was leaning around had a fancy hairdo and a low-cut dress that was open in the back. As I leaned around her one piece of meat was knocked off the platter by her hair and fell down inside the gap in the back of her dress. Since there were only eight pieces of meat for eight people I knew that I had to get it back, and I also knew that she was in trouble, so I quickly reached down her back, retrieved the wayward piece of meat, put it back on the platter, and continued serving. When I danced with her later that evening I thanked her for not making a fuss, and she thanked me for acting so quickly that she did not have time to do anything but gasp and act as though nothing had happened.

Although there wasn't a hockey rink in Fairbanks, an area on the Chena River in front of town would be marked off and flooded for games. A small area on the campus was also flooded for skaters and hockey practice. There were games during the winter between a Fairbanks team and the college team; then at winter carnival time a team from Dawson in Yukon Territory would come over for a series of games. Usually the Dawson team was the winner, but in the spring of 1936 the college team won the series. In the euphoria of winning, meetings were held and a decision was made to make up a team from the college, Dawson, and Fairbanks teams and head for the big time "outside," as we called the mainland United States. Doc Huffman, a dentist who was a great sports fan, organized games in southern Canada

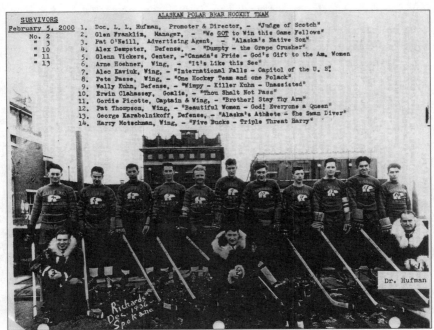

Patrick with Alaska Polar Bears hockey team in Spokane, WA, 1936.

and the northern United States. Glen Franklin was the business manager, and I was the athletic manager and handled publicity. We were able to use the college equipment and called our team the Alaska Polar Bears. Those of us from college took a year off from studies. We met the Dawson men in Vancouver, where we organized and had a few practices before playing a Vancouver team. We played so badly that the Dawson players left us, but we continued on to Seattle where we were able to pick up several players. We tied our first game in Seattle and won games in Portland and Spokane. We then went on to play in several cities in southern Canada, where we never won a game, and then in Minneapolis where we did win and chartered a bus to take us up to northern Minnesota, Michigan, and Wisconsin. We won or tied our games in those areas and made considerable efforts with publicity, like stopping traffic while holding powwows in our parkas and mukluks at busy intersections and having many news items in the papers. Despite all this, attendances were disappointing, and we finally went broke in Wisconsin. The bus was paid for back to Minneapolis so we all went there and then started scrambling to find our way back home. George Karabelnikoff wanted to see a girlfriend in Chicago and I wanted to visit uncles in Los Angeles, so we agreed to go together and each cabled to Fairbanks for 50 dollars. Mine came from the Chena Bar with a proviso that I work it off when I returned, and George got his from family. We went by bus to Chicago, where we had several fun days seeing the sights with his girlfriend. I was driving her car one day when a policeman pulled me over for making a wrong turn. He asked to see my license, which I could not produce, telling him that I had never had one as we did not need them in Alaska, especially for driving dog teams, which was what I usually drove. The policeman wasn't very happy with my answer and asked who owned the car. When Hilja, George's girlfriend, spoke up, he told her to take over the driving and not to ever let the dog team driver behind the wheel in Chicago again.

We saw an advertisement for men to drive Hudson cars and tow one from Detroit to Los Angeles. I managed to get a driver's license and we went to sign up. We were almost broke by then and you had to show 20 dollars when you signed up as the company would provide rooms but you had to buy your own food. I borrowed 20 dollars from Hilja and went in to sign up. Then I gave the 20 dollars to George so that he could sign up.

When he came out he started to give the money back to Hilja, but I asked her if we could borrow it and pay her back when we got back home, which she agreed to. We went by bus to Detroit; before we started out George and I pooled our money, which amounted to 24 dollars, so we each took half. The first overnight was in Indiana, where George lost eight dollars in a slot machine and gave me the two dollars he had left after dinner. By the time we got to Los Angeles we were eating mainly a bowl of chili with all the crackers we could get, and as our funds dwindled away one of us would get a cup of hot water and pour tomato ketchup in so we would have a cup of tomato soup. My uncle Wilson picked us up after we turned in the cars in Los Angeles and took us to his very nice home. They had a maid who washed our clothes, and we had wonderful food. Wilson let us use a spare car with a gasoline credit card in the glove compartment, so George and I drove around seeing what sights and museums we could see for free. My mother had three brothers, all in the movie industry in Hollywood. Wilson was a film specialist working mainly for Agfa. Fred, the oldest, was a production manager at Paramount. Charlie was a trick camera specialist responsible for the miniatures that would be suspended in front of a camera that would make it look like the scene was being shot in a beautiful ballroom or whatever. Fred would take us to prizefights in the evenings, where we met Spencer Tracy, George Raft, Edward G. Robinson, and others. He also arranged for us to get on some of the sets where they were filming. The one I enjoyed the most was one where they were filming Bing Crosby, Bob Hope, and Dorothy Lamour in one of the *Road to Rio* series. They had a few palm trees in front of a projected scene of a beautiful beach and ocean background. We met Bing, who had been a classmate of my oldest brother at Gonzaga University in Spokane many years before. When Fred was taking us back to Wilson's one evening, he asked what our plans were and I told him that George and I had been reading the ads looking for a way to get to Seattle but without any luck so far. We were confident that once we got to Seattle we could arrange to get back to Fairbanks. Fred asked how much money I had and I told him 15 cents. He then asked how much George had and I told him nothing since he was a big spender and I was holding our funds. He was quite annoyed that I had not told him of our predicament, but I replied that they had all been so nice to us that we were trying to resolve things ourselves. He said that he had access to cheap train tickets and would take us to the train the next morning, which he did. As he was saying good-bye he handed me 50 dollars and told me to send it to my mother when I got back to work as a gift from him, which I later did. We got off the train in San Francisco to see the city, since neither of us had been there. I located an old girlfriend from Cordova, and we had a great time as long as the 50 dollars lasted, and then we went on to Seattle. When we disembarked in Seattle I asked a cab driver how much he would charge to take us to the Savoy Hotel. He said that it would cost a dollar, and I told him that as that was all we had between the two of us there would be no tip, but he took us anyway.

As we walked into the lobby of the Savoy we saw Mr. Tibbets, who owned the Pioneer Hotel in Fairbanks and was one of the hockey team supporters. I said, "Hi, Tibs," and he growled, "How much do you two need to get home?" We asked if he could finance us for two nights in the Savoy and two steerage tickets on the next boat two days later, one for George to Juneau and one for me to Cordova; we could get family to help us from there. He did so but made us promise that we would put him high on our list for repayment when we got back to work. The day after I arrived in Cordova I saw Harold Gillam, who said that he had a flight to Fairbanks the next day and thought that he had room for me. After checking into Tibb's hotel I was walking down the street when Billy Root, the bus operator out to the mines, stopped his bus and said, "There's Pat O'Neill back from the hockey trip and undoubtedly broke like the rest of them," and handed me 20 dollars so I could eat until I got to work. I was soon working on a survey crew for the Fairbanks Exploration Company. It was still early in the spring, and we worked on snowshoes for several weeks to start with.

We ended up with substantial debts from the hockey trip. Although all had agreed that any profits or debts would be shared equally, no one except Glen and me paid off any debts, so besides the year of college we lost on the trip I had to stay out another year paying debts. I never had any regrets, though, as it was my first trip outside of Alaska and I saw and learned a great deal; it was a very worthwhile experience. When I was eating my first breakfast in Seattle, I ordered two soft-boiled eggs. They did not taste good, so I sent them back. When I told the waitress that the replacement eggs tasted just as bad she asked where I was from; when I told her that this was my first trip "outside" from Alaska, she said the trouble lay with me and not with the eggs, because I had probably never had a fresh egg before. She was right: the eggs we had in Alaska were all cold storage eggs that came up by boat and had been stored for quite awhile before they were eaten. There were 30 dozen in a crate, in layers of three dozen each. I once told someone that in order to extend the life of the eggs by preventing the yolks from sticking to the shell we had to give them a quarter turn each week. The person commented that it must have taken a long time to turn so many eggs and was embarrassed at how unthinking he had been when I explained that we turned them a case at a time.

There were some great people working their way through college. One that I greatly admired was Ivar Skarland, a Norwegian who was working at the Healy coal mine while he was learning English. Cap Lathrop, the owner of the mine, was a trustee of the college and took an interest in Ivar, and called President Bunnell to get Ivar into the college. While he was learning English Ivar was given work classifying prehistoric mammoth, mastadon, and other bones that were being collected from the nearby mining operations; thus began a lifelong interest in archeology and anthropology for him. Within two years of starting college Ivar was correcting papers for the English teacher. Ivar was older than the rest of us, and when a few of us would get together for an after-dinner bull session Ivar would sit in and say that the bull sessions were a good idea but he wanted to keep the talk on reasonably intellectual subjects. We had some great sessions.

In connection with his anthropology studies Ivar was measuring the heads of all the students. When he measured mine, he commented that it was an unusual shape, longer and narrower than the average. A few years later, Dr. Hooton, head of anthropology at Harvard, visited the college and Ivar acted as his escort. I was eating dinner at a restaurant counter in Fairbanks one evening when all of a sudden Ivar and Dr. Hooton appeared with their hands on my head commenting on the shape. What a surprise! One could never get annoyed with Ivar, though; he was such a charming man. Two or three years later when I was an aviation cadet at Kelly Field in Texas, we were told to report for head measurements so that average sizes for flying helmets could be determined. When I looked down the line I was surprised to see Ivar measuring the heads. When I arrived in front of him, he greeted me like a long-lost friend but then sent me on my way without measuring, as he said that I would throw the averages off.

Many of the classes were understandably quite small in the earlier years when there were very few students at the University of Alaska. In one geology class I had there was only one other student, Bob Wedemeier. Bob was a very bright, hard-working student. I worked hard but did not do as well as Bob. He graduated before I did and left Alaska shortly thereafter, and I did not hear anything about him until fairly recently. A former classmate told me that at a reunion on the campus Bob had asked if anyone had seen me in recent years. Our classmate told him that he had seen me often over the years and that I was president of International Mining Corporation in New York. Bob said, "My God, how did he ever make it?" Actually Bob was not the only one to be surprised, because I also wondered at times how I had made it from a very small village in Alaska to being president of a big board company in New York. From wearing snowshoes in Alaska to wearing wingtip shoes on Fifth Avenue was quite a jump. The same theme was probably on my mind while writing the commencement address I gave when receiving an honorary degree in 1976. The title I chose for my address was "Luck," which I described as that which happens when preparation meets opportunity—in other words, being prepared to accept opportunities when they develop. I have been fortunate throughout my life in being prepared and ready to accept possibilities when they came along. I went on to say how fortunate I had been, starting with getting a job in a mine when I was 15. I look upon that event as my first lucky one as it started me on my life's work. I was prepared mentally to accept a new direction in my life, and I was physically prepared to work as a laborer, which I did for several summers while working as a janitor during the winters to get an education at the University of Alaska. I then worked in various capacities in the mining industry, from laborer to engineer. During the war I spent four years as an officer in the Army Air Corps, where I learned additional managerial and organizational skills. After the war I progressed from engineer to superintendent while also gaining experience in many volunteer activities such as the Pioneers of Alaska, of which I was president; I was also chairman of the big celebration for the 50th anniversary of the discovery of gold in the Fairbanks area. So I was as well prepared as I could be when the opportunity came along to go to Colombia as chief engineer of a mining operation, which led to a position as vice president at the company headquarters in New York. A few years later I became president of International Mining as well as president or chairman of several subsidiaries or affiliated companies and director at various times of 14 or more other mining companies. I have always been very grateful for the opportunity to get an education at the University of Alaska, which along with work and other experiences prepared me for the opportunities that developed. People think about luck in different ways; I remember Jimmy Doolittle's comment, regarding his flying career, that luck is something that comes after you have taken every precaution to avoid the need for luck. I often think about being in the right place at the right time as being lucky, especially in regard to meeting my wife, Sandra, which was undoubtedly my luckiest day.

During my time at the University of Alaska the classes were mostly very small and the professors were very good. We did have one professor in practical mining, a big fellow we called "Skook" who wasn't great in the classroom, so the five of us students got together and planned our program of study, meeting one or two evenings a week, which worked out very well. Skook was, however, very good with the practical side of our studies and taught us to be miners while we were drilling, blasting, and mucking out an adit as an opening to bring power and heat from a power plant that was to be constructed on the flats below the college up to the campus on the hill. Skook also took us out to old mines in the area, where we learned underground surveying. One Friday afternoon as we were walking down a hill out in the woods to an old cabin where we were to stay for the weekend of work underground, we saw a moose close to the cabin. Skook was carrying his rifle as it was hunting season and killed the animal. He taught us how to skin and butcher it and hang it up. He took the liver for our dinner that night and it was very good. I had never cared much for liver before then but have liked it ever since; I learned that it is the only part of a freshly killed animal to eat without hanging. A year after the two-year course in practical mining I was working as a laborer in a bull gang on the creeks when we were called out in the middle of the night because a tunnel that brought the water for the mining operations had caved in and had to be opened as quickly as possible. I was at the face in the tunnel shoveling dirt into a mine car, throwing it over my shoulder as we had been taught and doing it rapidly. One of the big bosses had come to see how the work was going, and I heard him ask the foreman who the fellow was who was working so fast and well in the face. When he was told that I was an engineering student at the university, that was my opening to get off the labor gang and into the engineering office the next season. I liked the accounting classes that we had to take and one day mentioned my interest in business to the dean. He said that I would lose a lot of time if I changed to business and suggested that I stick with mining and take as many business courses as I could and perhaps end up in the business end of mining, which is what happened. While I was working for the U.S. Smelting Company after the war, I thought that studying law would be helpful in my mining career and started a home-study course with a lawyer in Fairbanks. I found it very interesting but only worked at it a year before the opportunity came along to go to Colombia. I was often sorry during my years in charge of operations in South America that I did not have the knowledge of law, as it seemed that a large part of my time was spent on legal and labor problems. However, I learned a lot from the many lawyers worked for the companies.

I have stayed active in one way or another at the university and for many years served on an advisory committee for the School of Mineral Industry. The second president of the university, Dr. Terris Moore, was a mountaineer and pilot and had been in Alaska off and on for over 30 years. This made him eligible to join the Pioneers of Alaska, which he did in 1952 when I was president of that organization, and I signed his lifetime membership card. He often said that he returned the favor in 1953, when he signed my diploma for the Engineer of Mines degree. Later on when I was making frequent visits to Boston as chairman of the Joslin Diabetes Center, I often visited Terris and his wife Katrina in Cambridge. Whenever anyone

Patrick at the University of Alaska in 1976 where he gave the commencement address and received an honorary degree. Rusty Heurlin (left) and Sandra (center).

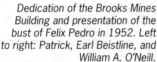

Dedication of the Brooks Mines Building and presentation of the bust of Felix Pedro in 1952. Left to right: Patrick, Earl Beistline, and William A. O'Neill.

from Alaska visited they would hoist the Alaska flag, which was always a nice touch to the visit.

My older brother Bill was on the board of regents for 25 years and was president of the board for the last five years. The trustees had been planning to give both Bill and me honorary degrees when he retired in 1975, but in late 1974 Bill drowned while on a consulting assignment in Liberia. They named a new research building for him, and I was honored to give the dedication speech for the William A. O'Neill building. The following year, 1976, I was given an honorary Doctor of Science degree and was the commencement speaker.

I was at the university in May 1973 to attend meetings of the School of Mineral Industry around the time of the annual commencement. One evening at a cocktail party the university president, Bill Wood, told me that he had arranged with the oil company ARCO to go to Prudhoe Bay the next day with James Michener and his wife, Mari, but that he would not be able to go because the regents meeting had not completed their agenda. He asked me if I could go in his place to accompany Michener and his wife. I was delighted with the invitation as I had read and greatly enjoyed many of Michener's books and was enthusiastic about the opportunity to spend some time with him. We left early the next morning with clear weather, so I was able to tell the Micheners about the places we were flying over as I had traveled much of it years before on exploration trips. At Prudhoe Bay we were outfitted with warm clothing, boots, and hard hats, and we visited the drills, shops, hospital, and other facilities including the pumping station at the start of the pipeline that transports the oil across Alaska to the ocean port at Valdez. We had lunch with the manager and several of the crew, among whom were men who had worked in oil fields all over the world in such places as Saudi Arabia, Kuwait, Iran, Iraq, Afghanistan, Egypt, Mexico, and the like. With every place they mentioned they were surprised to find that Michener knew more about the area and specific features than they did. It was absolutely amazing the detailed knowledge that he had of every part of the world they mentioned, and it was a most interesting three-hour luncheon.

Michener gave an excellent commencement address the next day at the university, and the day after that I took him and his wife out to see some of the mining areas where I had worked. We stopped to visit with Rusty Heurlin at Ester; Rusty was a well-known Alaskan artist I had known for many years. He had painted a series of large paintings on the history of Alaska, starting with the discovery of Alaska by Vitus Bering. I had helped Rusty mount the paintings on hangers for sliding back and forth while the story about each one was being narrated by Reuben Gaines. It is a most impressive series of paintings, each one about five feet by nine feet and

James Michener and wife eye pipe   Bit digs 1,400 feet down

VISITORS TO BP DRILLING SITE—British Petroleum's Charles Towill, left, leads author James Michener and his wife, at right, on a tour of BP's new drilling site on the North Slope. With the group, second from left, are Patrick O'Neill, Dr. Terrence Moore, former University of Alaska president; and Jerold Sorenson of UA.   (Staff photos)

British Petroleum's drilling site on the North Slope, 1972. James Michener and his wife (right); Patrick O'Neill (second from left).

extremely well done. The Micheners greatly enjoyed seeing the paintings and meeting Rusty and seeing his charming log cabins—one to live in and one to paint in. Michener and his Japanese wife were a charming couple, and it was an honor and a great pleasure to have been asked to show them around.

I had some prints of Rusty's arctic paintings when Sandra and I married. She had found it hard to believe that the wonderful pastel colors were truly representative, but when we made our first trip to Alaska together, as we were boarding a plane at midnight at Point Barrow in early July, there were those same beautiful colors on the horizon, so she knew they were real. The next evening in Fairbanks we had dinner with Rusty, and Sandra asked him to paint a large painting for us, which he did, of Eskimos and their dog team at first light in the Arctic after the 80 days of darkness. We cherish that painting and have enjoyed it for over 30 years. When Rusty was preparing to do the series on the history of Alaska he did small preliminary paintings of each of the proposed scenes on two- by four-foot pieces of plywood. The first time Sandra saw the large paintings she was enraptured with one of a dog team in the Arctic twilight. Much to our surprise Rusty autographed the small painting and sent it to Sandra, saying that he was destroying all the preliminary paintings but that as she was so excited about that one and we were so far away he had saved it for her.

Five or six years after my time with the Micheners in Alaska, Sandra and I and our children were walking out to catch a plane at the San Francisco airport when I saw Michener approaching. I told my family and said that I doubted very much that he would remember me but that I was going to stop him so that they could meet him. Just as I was about to speak Michener said, "Well hello, Pat, it is so nice to see you again, I saw where you received an honorary degree at the University of Alaska two or three years after I did and that International Mining had been taken over by another company and that you are now with Rosario Resources." I could hardly believe that he remembered me and was flabbergasted that he remembered reading those things about me. It was like one older, very well-traveled oil driller said at Prudhoe Bay, "I doubt that anyone else ever had such a retentive memory."

Two of my former classmates, Earl Beistline and Ernie Wolff, sponsored a history of the mining school at the University of Alaska and engaged a writer, Leslie Noyes, to assemble the history. It took 13 years and was published in 2001 with the title *Rock Poker to Pay Dirt*. It is the story of nearly a century of college students gambling on an Alaska-based mineral education that paid off in successful careers. It tells of the many significant contributions that a small land-grant college in remote Alaska, founded by Congress in 1915 and opened in 1922 with six students in a couple of frame buildings on a hill four miles from Fairbanks and an appropriation of $40,000, has made and continues to make to the economy of Alaska. Earl Beistline and Ernie Wolff spent the greater part of their lives as students, faculty members, and administrators of the School of Mines and had extensive experience as miners, engineers, and consultants as well as owners and operators of mining properties. They also personally knew all of the mining school graduates and had followed their careers from the time they were struggling, often hungry, poor students until they were successful operators or executives in industry. Regarding the book title, rock poker, also called petrological pinochle, was a game invented by Dean Patty, the first dean of the School of Mines, to make learning mineral identification more interesting. To hit pay dirt in placer deposits is to discover profitable gold-bearing ground. As I look at it, when a student is successful in his career and his years of study have paid off with a good salary, he has hit pay dirt. The success stories of many of the students from the School of Mines are told in the book, and I was honored to have been one of them.

3 Army Air Corps

I HAVE had a great interest in flying from the time in 1931 when Harold Gillam picked up our high-school basketball team in Anchorage and flew us home to Cordova when the normal rail and boat transport systems were shut down due to severe winter storms. I later worked for Gillam between high school and college, when he was pioneering an air route up the Copper River Valley. I went on many trips with him and also serviced planes of many of the well-known bush pilots of Alaska when they flew into Cordova. Later, at college in Fairbanks, when I had some spare time I would hang around the airfield to lend a hand or just visit with the pilots. In 1940 the army started a Civilian Pilot Training Program in conjunction with the University of Alaska, so I enrolled and obtained a private pilot's license. Whenever I had some money I would rent a small plane and fly locally,

Patrick's favorite plane, a B-17, and his favorite rank, captain, 1944.

77

often taking someone up to see the midnight sun if they paid for the plane rental. (During June and July in Fairbanks the sun goes out of sight only for a short time, and at two or three thousand feet it can be seen at midnight.)

In anticipation of possible entry into World War II the United States instituted the draft, and my number came up in early 1941 when I was in college. I applied for the Army Air Corps, hoping that this would get me into the branch of the service in which I was primarily interested and also enable me to stay in college until graduation in May of that year. I was called in August 1941 along with my roommate Joe DaGrade. We went out for an interview at Ladd Field. One of the two officers that interviewed me put one foot up on the table and asked me what color socks he had on. I said, "brown," and then the other officer showed me a picture of a Petty Girl and asked what I thought about it. I said something to the effect that it was a beautiful, sexy girl and the officer said that I seemed quite normal, and I was accepted as an aviation cadet subject to passing a physical exam. I did not pass the physical because of several missing teeth and enlarged tonsils, so I was told to get a partial denture and have my tonsils removed and then check back for re-examination. Joe did pass the physical and left soon after for flying school but was killed in a training accident several months later. I reported back to Ladd Field in late September, passed the physical exam, and was told that I would be called soon. I did not hear anything until Sunday, December 7, Pearl Harbor Day. I had been to a late party Saturday night and was sleeping when a friend called to tell me about the surprise attack in Hawaii. While I was listening to the news I received a call from Ladd Field telling me to report at 8 a.m. the following day. By noon on Monday I was all signed in and received orders to leave on Pan Am Wednesday morning for Seattle and to go from there by train to Kelly Field in San Antonio, Texas. There wasn't much time to terminate my employment at the company, move out of my apartment, and ship my belongings to my parents' home in Anchorage. I was told to take just what I needed to get to Texas, as I would be outfitted with army clothes and would have to send my civilian clothes home or discard them. On the flight to Seattle there were eight of us, which was the capacity of the Pan Am Lockheed plane, all en route to Kelly Field. Two of the eight had also taken the Civilian Pilot Training Program and the others were enlisted men who had applied for flying school.

Upon arrival at Kelly Field we were each assigned to a specific squadron and issued G.I. (general issue) clothing, toiletries, bedding, a footlocker, and so forth. We were then sent for physical examinations and inoculations, had our hair cut very short, and started training in marching with daily calisthenics. We were taught how to make our beds properly and to organize our clothing and other belongings. The first time out for physical training after calisthenics we were told to get organized for football scrimmage. I was standing on the side trying to figure out what was going on when I was called over by the officer in charge, who demanded to know why I wasn't participating. He could not believe that I had never seen a football game and had no idea what to do or how to throw a football. However, I soon learned. The first evening we were allowed to go to the PX (post exchange), where we could purchase postcards and stationery to write home with our address. As each of us reached the cashier with our purchases she would ask where we were from. When I told her I was from Alaska, she said, "Ain't that one of those northern states up near Oklahoma?" I tried to briefly explain where Alaska was, but her knowledge of geography was practically nil and the line of cadets forced me to move on; really, I don't think she cared anyway.

The Military Science and Tactics classes I had taken in college helped me at Kelly Field as I was soon made a platoon leader, which made drilling more interesting at least. Upon completion of preflight basic training at Kelly Field the cadets were sent to various primary flying training schools, which had been set up rapidly to accommodate the influx of students, as our country was getting onto a war footing as rapidly as possible. I was in a group sent to Ballinger, Texas, where all the instructors were civilian pilots, and there were only three or four Air Corps officers and a few enlisted men who supervised our military training. The pilots were mostly experienced men and good instructors. The flying training I had previously was helpful, and I soloed very quickly. The hazing that went on at all the flying schools was in full force at Ballinger, so the first half of the time there we were hazed by the upperclassmen and then it was our turn to haze the underclassmen when they came in. Much of the hazing was demeaning and silly, but it was supposedly to teach us to take orders speedily and without questioning authority. We were training in PT-19s, single-engine, low-wing monoplanes with two open cockpits. The student was in the front cockpit with a mirror

Patrick as a pilot in World War II, 1942.

in which he could see the instructor, who had a speaking tube. As he gave instructions you would nod if you understood or shake your head if you did not. Instruction in aerobatics started soon after soloing. One instructor told a classmate of mine that he was going to teach him how to roll the plane: to start with he would fly upside down for a few minutes for the student to get the feel of it. He told the student that it was important to keep pressure on the pedals so that his feet would not drop down, as rudder action was essential in any aerobatics. He then told the student to check that his seat belt was secured, and when the student nodded to confirm that it was the instructor rolled the plane over and promptly fell out, as he had not secured his own seat belt. He opened his parachute and landed safely, although on a cactus plant, and was on his stomach in the hospital for two days while they were pulling cactus needles out of his rear end. The student landed the plane safely, and I went with him that evening to see the instructor. The man was very embarrassed, but we all laughed about it. One evening when we were sent out to practice aerobatics alone I saw the plane in the next area to mine flying very erratically. The student's head would disappear and the plane would level off; then the head would reappear and disappear again. Finally the student's head stayed up and he flew back to the base and landed. I landed soon after and asked him what had been going on, and he said that when he first rolled over he had pulled the control stick out and had had trouble getting it back in so that he could properly fly the plane. Late one day when I was returning to the base the wind had come up very strongly and the dust was blowing. I was landing into the wind but also into the setting sun onto what I thought was the gravel airstrip but which in reality was blowing sand. I stalled for the landing several feet above the runway and dropped like a rock. I thought that surely the wheel struts would come through the wings, but they didn't. The plane was all right, but I was certainly shaken up.

Ballinger was a small town without much in the way of nightlife, although the people there were very nice to us cadets and had small parties at some of their homes on several occasions. San Angelo was the closest city, and one Saturday night one of the fellows who had a car invited three of us to go with him to a nightclub there. There was an advanced flying school nearby, so there were many cadets trying to get a dance with the few girls who were there. There were four girls at a table near where we were standing, and I told my companions that the only way we were going to get a dance was to decide which girl we were going to ask and be there by the time the music was ready to start, which we did. The girl I asked started to unwind as she stood up, and she said, "Don't be so surprised stranger; I am just one of those long, tall Texas gals." She was too tall for me to see over her shoulder and not tall enough for me to see under her arm, so I asked her if she would like to lead. She said that she often led so off we went and had a good time and danced several times together.

I was co-editor of our class book at Ballinger. It was a lot of work but a nice thing to have for memories, and it was the only place to publish a class book and the only place where I had pictures of classmates in Air Corps training. Our group was assigned to different basic and later advanced flying schools, so we seldom saw each other again. However, during my four years in the Army Air Corps I would occasionally meet someone from our primary training group. The sad part was finding out about those who had been killed, of which there were many.

From primary training I was assigned to Randolph Field near San Antonio for basic training. Randolph had been the main flying school for the army for many years and was known as the West Point of the air. The facilities were excellent, and the barracks, although crowded to accommodate greater numbers of cadets, were very good. As an underclassman I had a cot in the hallway and shared a bathroom with five others, but when I became an upperclassman and cadet captain of company G, I moved into a room for three. The organization and discipline at Randolph were outstanding and very efficient. If the entire Air Corps had been managed as well, I would have been very happy to make a career in the Army Air Corps, but unfortunately I never saw another organization so well run the rest of my time in the service.

For any infraction of the rules you would get a demerit, or "gig" as it was called. Infractions included being late for or missing reveille; being late for any of the activities such as calisthenics, drilling, or classes; not being dressed properly; not having your bed and clothes in proper order; or many other things. For each demerit you would have to march back and forth on the ramp for one hour whenever you had free time (such as Sunday or occasionally Saturday afternoon, when one could go off the base into San Antonio). Everyone had to gather by groups for reveille and roll call at 6:30 every morning. There were two men from the Bolivian army who were receiving training along with us in company G, and they would try to get me to report them present at reveille so they could sleep later, but I told them that an officer frequently went through the barracks while we were standing reveille and that if I reported them present and the officer found them in bed I would be the one walking the ramp, so I would not take the chance. I did not see the men after basic training until many years later when I flew with them in Bolivia, which I tell about in the mining chapter.

We drilled every day and had a parade every Sunday. It was thrilling for me to lead company G on parade. The cadet captains had a sword, which they held in salute while passing the reviewing stand. Marching behind our U.S. flag with the band playing was a moving experience, and I was very proud to be in the military and leading a company of cadets. There were eight companies, and when we were lined up for my first parade as captain I thought it would be appropriate to say something to my group so that they would do their best. I had heard the captains of adjoining companies threatening their men that if anyone got out of step or out of line they would be given demerits and be spending their free time walking the ramp. I thought that if I was in one of the other groups I would not like to be threatened, so I told my group, "Whichever company wins the parade will be excused from drill practice for three days, and I'm sure that we would all be happy to miss some drill practice, so let's go out and do our best." It worked, and our company won first place in several parades.

At Randolph we were training in old BT-9s, which were well known as being difficult to get out of a spin, but we were required to put them into spins as well as doing other aerobatics. Two of my group were on a buddy practice session one day and were unable to get the plane out of a spin so they bailed out. One of them pulled his ripcord too soon, and a line on his parachute caught on the tail of the plane and he was killed. He was a very well liked fellow from Georgia and had worked with me on the class book we put out at Ballinger. As cadet captain of our company I sent a telegram of condolence to his family, telling them of the profound respect and admiration we had for their son. I added that his memory would serve as an unlimited source of inspiration to all of his classmates. He was the first of our class to be killed in training, and it was a very sobering and sad time for all of us.

Not long after I was named cadet captain I was summoned to a meeting with one of the senior officers. He talked about loyalty to our country and the great importance of being alert to any indications of subversion or disloyal acts by anyone and said that I should always keep my eyes and ears open and report anything that I ever saw or heard that was suspicious or possibly harmful to our country or our troops. Certainly there were spies around, as we well knew from groups like the B-26 group, which I was in for awhile. This group, as it was approaching the north coast of Britain on its flight from Newfoundland, was met by German fighter planes; we heard that almost half of the group had been shot down. Someone had found out about the plans of that group. While I was in the service I was contacted at each new location and advised whom to contact if I had any information. During the 30 years that I was active in management in South America and other countries I was also contacted and interrogated frequently by CIA agents about activities of foreign agents in our labor unions in South America.

From Randolph Field I was transferred for advanced training to Brooks Field, which was on the southwest side of San Antonio. I was cadet commander of the entire group of aviation cadets at Brooks, which took a lot of time but carried a few perks like a private room in the barracks. We flew

AT-6s, which had retractable landing gear and were much faster than the previous planes in which we had trained. We learned formation flying as well as strafing and combat fighting and sometimes would have fun following each other through the beautiful cloud formations in that part of the country. We practiced short-field landings and takeoffs in farming areas with trees on the approaches. One area in a large cornfield had been cleared for our practice. The instructor would sit in his plane near the end of the cleared area and give us instructions such as to go around if we were coming in too fast or too high; when we landed, we would wait near his plane until he cleared us to taxi back to takeoff. One classmate named Gene had to go around once and then ignored the instructor's advice to go around again; he landed too far down the strip and went past the cleared area into the cornfield. He was cutting corn and throwing it into the air while the instructor was telling him to retrace his path and get back into the cleared area; the instructor was furious and gave him quite a tongue-lashing. A few days after the cornfield incident Gene was landing after me at another small field we were using and was on his final approach with his gear still up. There was a horn right behind your head that blew when you cut back on the throttle for landing if you did not have the landing gear down; the instructor was hollering into the radio for him to go around and get his gear down but Gene kept on coming and landed with his wheels up, damaging the plane extensively. I was parked close to the instructor's plane and cut my engine and went over with the instructor to Gene's plane. The instructor asked, "Didn't you hear me telling you to go around and get your gear down?" and Gene replied that the horn was making so much noise he could not tell what the instructor was saying. That was the end of Gene's days as an aviation cadet. We had quite a few cross-country flights, and the ones at night were a bit scary at first, as our only navigation aids were occasional flashing beacons and you had to be confident with Morse code to know where you were.

Graduation at Brooks was on August 5, 1942, six days before my 27th birthday. At that time age 26 was the deadline for getting your wings, so I was just under the wire. At the graduation ceremony the commanding officer handed us our second lieutenant bars and our wings, and when the ceremony was over family members or hometown girlfriends who had come for the graduation would pin on the bars and wings. However, the only people I knew were my classmates, and they were all busy with family and friends. I felt awkward standing alone in the crowd trying to put on my bars and wings, so I sadly walked behind the hangar and was trying, not very successfully, to put them on myself. While I was struggling an older master sergeant who was in charge of maintenance on the planes we had been flying walked up and said that he was sure I would not have any family from Alaska there. He added that he admired the way I had done things as cadet commander and that he would be very proud if he could pin on my bars and wings. He said that he wished more than anything that he could someday do the same for his own son. I was very grateful, and we hugged and I went back to the celebration.

My orders were to report to Selfridge Field in Michigan to join a P-39 fighter group that was in final training before going to England. A classmate, Maurice Boggess, was also being sent to Selfridge Field. He had a car and invited me to ride with him if I paid for the gas. We stopped to visit his family and his sweetheart in Iowa, who flew out to Michigan a few days after we arrived there. Maurice asked me to be best man at their wedding, which I readily agreed to do. It was an early-morning wedding at the Catholic Church in the nearby town. Doris stayed at a motel nearby and had Maurice's car, and we were to meet at the church at 7:45 in the morning. When Doris was not there by 7:50 I went to the motel and woke her up. She was flustered and nervous, and as soon as she got the basics on I helped her and we dashed to the church. The priest had held the Mass but Maurice was an emotional wreck. He was on the verge of fainting, and I practically had to hold him up during part of the service. However, it was a very successful marriage. I kept in touch with them at Christmastime over the years. Maurice had a sand and gravel business after the war, and they had nine children. I was invited out to their 50th wedding anniversary in 1992 but unfortunately could not go due to illness of my mother-in-law. I did go out for Doris's funeral three years later; Maurice died shortly after that.

Like every new pilot who will never really be as good as he thought he was when he graduated from flying school, I was eagerly looking forward to getting into combat and was pleased with my assignment with the P-39 Group. I was soon disappointed, however, when the commanding officer called me in and said that he had permission to assign me to his staff on the base and that I was to be the base technical inspector, maintenance officer, and assistant operations officer. It was really a disappointment, but I was too busy to worry much about my fate. I later became friendly with the personnel officer and asked if she had any idea why I had been selected from the new group to stay on the base. She told me that she had reviewed the files with the base commander on the eight pilots who had arrived to join the P-39 group and that two letters of reference in my file were so outstanding that I was selected. She said that one letter from the president of the University of Alaska said what an outstanding student I had been, an exceptionally bright engineer, a real leader in student affairs, an athletic manager, and so on. The other letter she referred to was from Jim Crawford, manager of the U.S. Smelting, Refining, and Mining Company in Fairbanks, for whom I had worked in several different capacities. According to the personnel officer Jim was very effusive in the letter, saying what an industrious, hardworking engineer I was, very intelligent, adaptable, dedicated, loyal, and so forth. I was modest enough to know that the writers of the letters had stretched the truth considerably in trying to help me with my career in the military, and indeed they did help me. Despite my disappointment in not going off to combat I did gain great experience and management skills during my four years in the service that greatly helped my career in later life. It was several years before I fully realized just how much President Bunnell and Jim Crawford had been carried away in their praise during the country's early-wartime euphoria. I have already mentioned how President Bunnell referred to me a bit backhandedly as "the best janitor we ever had." As for Jim Crawford, four years after I was named dredge superintendent I decided to accept a position as chief engineer at a mining company in Colombia, and Crawford was not at all pleased with my decision. I had hoped that he would understand that I was leaving because of the limited future of the company in Fairbanks, but after I was transferred to the head office in New York I was able to look at my personal file and in it was a letter from Crawford, which stood in sharp contrast to the very good recommendation he had given when I was entering the service.

My roommate for some time at Selfridge Field was Joe Murphy, who had been in the service in the First World War and had come back in as a captain. He was from Louisville, Kentucky, a self-styled Kentucky colonel. Whiskey was hard to come by during the war, but Joe regularly received a case of bourbon, and we never lacked for friends to help us drink it. Joe also liked to fly, and there was a small single-engine Aeronca plane available, so I would take him out occasionally in the evening to sightsee around Gross Pointe (the luxury residential area of Detroit) and to do some aerobatics, which he enjoyed. A nurse heard Joe talking about our flights and asked me to take her up. When I did a loop, she started screaming and asked me to take her back. She had on light tan slacks and was quite a sight as she walked away from the plane with the seat of her pants all wet. Joe had friends from Kentucky who had a sugar beet farm in central Michigan. They called Joe when harvesting was finished and said that we could land close to their house, so one Sunday Joe and I went there in the Aeronca for lunch. They had a very comfortable home and served great mint juleps before lunch. I could not resist the mint juleps but had to sleep for three hours before I dared fly back to Selfridge. That was the only time I ever flew a plane after drinking, and even though I felt OK after sleeping I knew it was wrong, and I never did it again.

One time several planes of a B-17 group stopped overnight at Selfridge Field en route overseas. One of the planes was carrying Clark Gable, and it ran off the runway on landing and damaged the outboard engine and a wing tip, so he was at Selfridge for three days waiting for a replacement plane to arrive. We had a big party in our room one night and Clark got very drunk and then very sick. Joe was helping him out of the bathroom and into bed when he asked me to check the toilet as he thought that Clark had lost his false teeth. I found his upper plate and Gable could not have been more appreciative the next day.

Joe commanded a group of soldiers, and one day he called and asked me to come over to his office. A man in his group had died and they could not locate any relatives. All he had was 64 dollars, so he was to be buried in a pauper's grave with all his possessions. The sergeant in charge of personnel suggested that someone write a check to the deceased and that the squad use the cash for a wake for the man. As commanding officer Joe did not think it a good idea for him to write the check, so he asked me to do it, which I did.

One day as Joe and I were walking back to work after lunch a soldier approached us and said that he had just received orders to ship out overseas the next day. He had a car that he needed to sell and he needed the money that day so that his wife could travel home. I did not have a car but was willing to acquire one, so we looked at the car, a 1940 Chevrolet in good condition. We agreed on a price of six hundred dollars, but the problem was how to get the cash that day. My bank account was still in San Antonio and Joe's was in Louisville, so we went to the local bank in Mount Clemens and asked to see the manager. We explained the situation to him, and Joe said that the head of the bank in Louisville was a personal friend and asked the manager to call and get authorization to cash a check for six hundred dollars. The manager did so and put Joe on for a few minutes to verify that it was him, and then he got the authorization and was smiling when he hung up. He told us that the banker in Louisville had not only approved the six hundred dollars but had told him that if Joe wanted to buy his bank he would also approve that.

After repairs were completed on the B-17 that Gable was in, it had to be test-flown before it could be released. No one at Selfridge had ever flown a B-17, so the operations officer, Al Daily, and I read the technical manuals and taxied the plane around and then took off. Al made the first flight and I made the second, so we checked each other out in a B-17. I flew them extensively later on in Amarillo.

The colonel who was commanding Selfridge Field was a very heavy drinker, and one evening when he had had more than he should have he was going to his office. He had his young son with him, as well as his driver, and as the boy ran ahead up the stairs the driver caught up with him and picked him up. The colonel pulled his gun and told the driver to put the boy down and then shot the man. The colonel was court-martialed and I was one of the many officers called to testify. I had flown with the colonel on several occasions and was asked about unreasonable rage, intolerance (since the driver was a black man), unusual behavior, or excessive drinking, and it was mainly the drinking that was his downfall. He received a short jail sentence and was discharged from the service.

I had requested an overseas transfer, either to a group or as a replacement pilot, and was sent to Fort Wayne, Indiana, to a B-26 group that was in the final stage of training before going overseas. The planes were the early Martin B-26s with a round fuselage, two high-powered engines, and very small wings. The wings were so small that the plane was called a widow maker or a flying prostitute because of very little visible means of support. A significant number of the planes and personnel had been lost in training because of a sudden loss of power in one or the other engine on takeoff, when the plane would roll toward the bad engine and crash. Once it was up to speed, the plane could be flown with one engine but would not fly with one engine at low or takeoff speeds. The only solution when one engine failed on takeoff was to immediately cut the other engine and crash straight ahead. Although the plane was usually demolished, most of the crews survived. Because of my experience as technical inspector and maintenance officer at Selfridge I was pulled out of the group and assigned to the base with the charge of learning the reason for so many engine failures. Along with another officer and two experienced master sergeant mechanics we started testing any carburetors we could recover from crashes. We also started bench-testing new carburetors in the maintenance shop and at Wright Field in Dayton, Ohio, where the top technicians and factory representatives were located. It was soon discovered that a small rubber part in the carburetor had a tendency to collapse under maximum power, shutting off fuel flow. Replacement with a stronger rubber part solved the problem. By the time we completed the test work on the bench tests and in the air the group had left for England, so I was ordered back to Selfridge. However, while I was at Wright Field I saw Sir Hubert Wilkins, a famous arctic explorer, who was advising the army on cold-weather equipment and activities in the Arctic. I had seen him once in Alaska when he visited the university, and I spoke to him about the work he was doing. He said that he thought that I should be involved where my experience in the Arctic would be useful and that he would see what he could do. Not long after I returned to Selfridge I was ordered to an arctic search and rescue group at Buckley Field near Denver, Colorado. When I checked in with the personnel officer at Buckley I was told that the commanding officer wanted to see me and I thought, here we go again, and sure enough he said that they needed someone with arctic experience and that he was assigning me as an instructor to the arctic search and rescue group. He would not listen to what I thought was logical reasoning as to why I could do more actually in the Arctic with the group, so I spent several months as an instructor. In teaching pattern searching one of the instructors would fly out early and land in a small field on a farm or anywhere there was a small, cleared area and wait to be found. One day I was about halfway to the area where I was supposed to go when my engine quit. There was only one very small area that I could see where I thought I might be able to land, so I circled and landed. However, the field was too short, and I went through a fence at the end of it, damaging the propeller and scraping up the plane in addition to damaging some of the farmer's crops. I set out for the farmhouse a mile or so away but met the farmer on the way. He said that he could take down some fence so that another plane could land with a mechanic and a new propeller, enabling me to take off after repairs. I told the base where I was so that the search group would not spend too much time looking for me in the original search area and arranged for a mechanic to come out. Meanwhile the search pattern was enlarged, and one of the planes finally spotted me, so the day wasn't completely lost. My plane was fixed up, and I was back at the base in time for a late dinner.

After the training was completed for the arctic search and rescue group, I was transferred to Amarillo Army Air Base at Amarillo, Texas. At my interview with the personnel officer I mentioned again my desire for overseas assignment and soon after was transferred to an air sea search and rescue group in Biloxi, Mississippi, that was scheduled to go to Southeast Asia. The group was just getting organized and waiting for more planes and personnel. They had only one PBY Catalina amphibian plane, which we were using for training. However, we soon received word that the Royal Air Force had decided to take over air sea rescue in the area to which we were being assigned, so our group was disbanded. I was then assigned to Valdosta, Georgia, as an instructor in a twin-engine advanced flying school. The most pleasant part of that assignment that I recall occurred while I was driving from Biloxi to Georgia. Somewhere along the way I saw a sign by the road—three cantaloupes for 50 cents—which sounded like a great bargain to me: cantaloupes were so expensive in Alaska that I rarely had a slice of one. I stopped and bought three, borrowed a spoon, and ate all three on the spot. They were delicious, and I always remember that stop with pleasure. It was very awkward for me at the Valdosta school as I was a captain by then and outranked most of the staff as well as all of the instructors. I got in contact with the personnel officer back at Amarillo, who told me that they had a need for an aircraft maintenance officer and that he would request me. I was soon back at Amarillo with much responsibility and very interesting work as the base maintenance officer. Amarillo was a stopover for cross-country flights, mostly for planes manufactured on the West Coast flying to the East for delivery overseas, so we had over 50 planes of many different types stopping for service every day. There were also 10 to 15 B-17s stationed there for training mechanics and crew chiefs. The school for training mechanics was excellent and trained many mechanics in a relatively short time; the men who came from farming areas did especially well as they already knew a great deal about handling tools and repairing equipment. There was a production line maintenance setup where the mechanics would work progressively from the tail to the front of the plane, including the engines and electrical and radio equipment. As soon as the supervisor certified them, they would go out on the line as crew chiefs and from there would go overseas as replacements. It was inspiring to see the men develop in ability and responsibility. After considerable experience with the B-17s I was designated an engineering test pilot and had to test fly every plane as it came out of maintenance. To start with I had some problems with careless work such as cowling falling off, engine failures, an engine falling off, another engine catching fire, and so forth. At a meeting with the commanding general about the problems I suggested that all the mechanics on the production line be put on one-half flying pay so that they would have to fly when ordered to do so. When I announced this the next morning, there was great elation, which quickly changed when I told them that I would pick out five or six of them at random each day to accompany me on my test flights, and I read off the names of the five who were to fly with me that day. The two planes I was scheduled to fly that day were suddenly scratched. From then on I never had any trouble with the planes after maintenance.

Occasionally some high-ranking officers would stop for refueling; as soon as we found out that VIPs were on board we would call the commanding general's office, and he would dash over to pay his respects and be sure that we were doing everything possible to get their plane serviced and on its way. There were planes coming and going around the clock, so I would often stop late in the evening to check on activities. Late one night when I was at the operations office the tower called down to say that a VIP was landing but that he had said not to call the base commander. I met the plane, and it was carrying Lt. General Joseph, known as "Vinegar Joe" Stilwell, commander of all troops in the China–Burma area. I said that I had standing orders to alert the base commander whenever any VIP landed. He asked how many stars the base commander had, and when I said, "one," General Stilwell said, "Well, I have three stars, so you can tell your boss that I ordered you not to contact him." The general said that he really wanted to get a bite to eat and some sandwiches for his crew while the plane was being serviced and get on his way to Washington as soon as possible. The only food place open at that time of night was a coffee shop across the field where commercial flights operated, so I alerted the tower, and the general and I got into a jeep and drove across the field and picked up some food. He was very interesting and a delight to be with. I was called on the carpet the next morning when the general found out about Stilwell's visit, but he wasn't too rough on me after I told him what Stilwell had said about having three stars.

I had a wonderful secretary at Amarillo, Private First Class Reno Zancanella, who had left his business in Wisconsin to enlist and do what he could in the war effort. He was about 15 years older than I and handled my office very efficiently and kept me out of trouble. I would often be so busy that I would not go to calisthenics, which we were all required to attend. The first time I missed them without a prior excuse a letter came from headquarters wanting an immediate explanation. Reno said that no matter what I wrote back I would be in trouble, so the best thing to do was to throw the letter into the wastebasket as the head office usually did not follow up. I did that from then on whenever I received letters about unimportant matters and never got into trouble about them. Reno and I kept in touch after the war until he died recently at age 97.

I have already described the Gray Ladies, a wonderful group of elderly women who set up a place at the operations office at many of the airfields around the country. They would serve sandwiches and coffee to crews while their planes were being refueled so that the crews could eat while they were filling out their flight plans and then get on their way. It saved a great deal of time and greatly helped in keeping planes moving. I have also mentioned Mrs. May Wells, who was short and gray-haired and looked very much like my mother. Only the in-transit pilots were supposed to be fed by the ladies, but I was around the operations all day, and Mrs. Wells made an exception for me, so I had a sandwich there regularly and seldom took the time to go to the officer's club for lunch. This is what led to the lifelong friendship between Mrs. Wells and my mother.

The two tech sergeants who were in charge of the crews who parked and serviced the many planes that came and went were excellent men and very dedicated, conscientious workers. I was quite demanding at times and kept on top of everything that was going on. I never realized how hard I was on the men, though, until one very windy day I was walking into the wind as I came to the corner of the hangar and stopped right behind the two sergeants just in time to hear one of them saying that he would like to be a major for half a day, because it would give him great pleasure to chew the ass out of Captain O'Neill. The other man said, "Hell, you couldn't even sink your teeth into that iron-ass bastard." At that point I said, "Hi," and asked what they were doing standing around talking when there was so much work to be done. They were shocked and surprised and said that they had been discussing work assignments for their crews; I tried to be a bit smoother in my demands after that. Several years after the war one of the sergeants made a trip by camper to Alaska and inquired about me from the listing for one of my brothers in the Anchorage phone book. I happened to be in Alaska at the time, on vacation with my family, so he caught up with us and we had a nice visit—so good, it seemed, that he decided to travel with us and followed us for several days. So I guess I wasn't too bad to work for.

One time the executive officer of a B-17 group in England passed through Amarillo, and I asked him if he could use another pilot who was an experienced maintenance officer. Shortly afterwards I was called to the office and told that they had a request for me, but that orders had come in for me to attend Command and General Staff School in Leavenworth, Kansas, and that I had to go there. I was very disappointed at missing another opportunity to go overseas but must admit that the school was excellent and a great help in later business organization and management. It would have been essential if I had wanted to stay in the service, which I was not yet sure about at that time.

The school had excellent instructors, most of whom had combat experience and knew what they were talking about. It was a tough school with frequent exams and if you did not have reasonable grades you were expelled. Also the commanding general would walk around often, and if he caught anyone napping he would expel him right on the spot. The days were long and homework extensive, but it was well worth it.

A few B-17s that had been damaged in combat but which were still flyable were returned to the States, repaired, and used for training such as that at Amarillo. We received a badly damaged one of these, and the maintenance people stripped it of its turrets and patched it up, and we put the name *Old Patch* on it, and often used it for trips around the country. One general we had at Amarillo for awhile had one lady friend in Tampa, Florida, and another in Los Angeles, so we had training flights to one or the other of those places occasionally. I learned to play bridge, and we tried to have enough bridge players on the long trips to have a fourth and play a good part of the time. One time another pilot and I had to take some people to Boston in *Old Patch*, and as usual we would take along people who were traveling. On our way back we had very bad weather the last hour or so before we reached Amarillo. One of the crew came up to the flight deck and said that the passengers were so scared they were all putting on their parachutes. I looked back and saw one of our crew helping one of the WAACs into a parachute. Since she was wearing a skirt I thought the men were just having fun, but sent the crew chief back to tell the passengers that although the weather was bad we were not in any danger, adding that although it was always prudent to wear a parachute it was not necessary at the time. However, not long afterward it felt like something hit the tail area. The crew chief ran back and found that one of the men had jettisoned the door and bailed out. He sent a man up to tell me what had happened and stayed by the door to prevent anyone else from bailing out. The door had hit the tail assembly, which is what I had felt. We called the base at Amarillo to report what had happened and to give our position. By the time we landed about 40 minutes later the soldier who had bailed out had phoned the base to say that he had landed safely and would look for transportation to get to the base. He was disciplined and had to spend some time with the base psychiatrist. Other than having to write a detailed report I did not have any trouble over the incident.

A major who was with a B-29 group that was almost ready to go to the South Pacific stopped at Amarillo in late 1945, and I had a chance to talk to him about my joining the group. He was impressed with our maintenance organization and said that he would put through a request for me to join his group. About ten days later a request came through. However, when the personnel officer processed the papers for me to go to B-29 transition training prior to joining the group, the papers came back stating that one had to have completed B-17 transition prior to going to B-29 transition and that my records did not show where I had ever been through B-17 transition. We sent my records showing that I had been checked out in a B-17 at Selfridge Field and had over 200 hours as an engineering test pilot on B-17s, but whoever was processing the papers was a stickler on regulations, and the order came through for me to report to B-17 transition. So I went to Hobbs Field in New Mexico, and as soon as I had had a check by a senior instructor I applied again to go to the B-29 group, but was assigned as an instructor on B-17s in the meantime. The personnel offices dragged their feet on my requests so long that the war ended while I was still instructing.

Several men that I knew wanted to get in the war before the United States entered, so they went to Canada or England and joined the air services there. One of them became a pilot in the RAF, and after the United States entered the war all of the Americans were offered a transfer to the U.S. services or otherwise would lose their U.S. citizenship. The man I knew decided to stay with his RAF buddies. He was shot down over Germany and broke both his legs on landing and spent the rest of the war in internment camps. His legs had not been set properly, so after the war he applied to get his citizenship back and came to New York where his legs were re-set. He was a great party man and one night he went out with a group at a club in New York. While everyone was away from the table, he passed out and slid under the table. When some other people sat down, one young lady hit something under the table and looked down and saw my friend. They pulled him up, and he joined their party. A year or so later he married the girl that found him under the table.

I had seriously thought about staying in the service. Besides doing well in all my classes and at the Command and General Staff School, I had excellent reports and recommendations from every senior officer I had worked under. Nevertheless, I was so disappointed with the red tape that prevented my going overseas that I decided to get out as quickly as I could. I had spent eight years studying mining and only four years in the Air Corps, so going back to mining was really a more logical decision. I have always been very glad that I did go back to mining as I have been actively involved in mining to past 80 years of age whereas my flying days would have ended at age 60. I went back to the U.S. Smelting, Refining, and Mining Company in Fairbanks as exploration engineer at a salary of 400 dollars per month. This was less than I had been making as a captain on flying pay, but I was not too concerned as I was confident there would be opportunities for advancement in the company. However, I was sorely tempted by Bob Reeve, whom I had helped out in Cordova when he first came to Alaska, as he was getting started with a flying service. He had done considerable contract flying for the army during the war, especially to the Aleutian Islands, and was organizing regular service to the area with Reeve Aleutian Airways. He came to see me on three different occasions; he started out offering me 500 dollars per month to work for him but by the third visit he was up to 1,000 dollars per month. It was hard to do, but I asked him to please not tempt me anymore, as I had made up my mind to return to mining and I intended to stay in mining. About 12 years later I was sitting beside Bob at an Explorer's Club function in New York, where I was by then executive vice president of international mining, and he said that as much as he would have liked me in his organization I had certainly made the right decision to stay in mining and that he admired my determination. I always stopped in to see Bob Reeve whenever I was in Alaska and had many great visits with him. He received an honorary degree from the University of Alaska, and I never saw anyone as proud as he was. On one trip I was invited by my brother, Bill, who was president of the university board of regents at the time, to a dinner in Anchorage with General Jimmy Doolittle (of the famous Doolittle Raiders that made the first attack on Japan with B-25s flying off aircraft carriers). Also at the dinner were Doolittle's wife, Bob Reeve and his wife, and Lowell Thomas. The evening was very interesting and lots of fun, and went on until early in the morning. Doolittle, who had just received an honorary degree from the university, was a great storyteller, and his wife would suggest that he tell this or that story whenever there was a lull in the conversation. Bob Reeve was a great storyteller himself, and Lowell Thomas had traveled to more places in the world than anyone and could tell about his travels and incidents in an intriguing and interesting manner, as he did in his regular broadcasts for many years. When I was working for Harold Gillam in Cordova, Bob Reeve had recently arrived from flying in South

America; Gillam could not offer him a job but suggested to Bob that there was a plane not being used in Valdez, so Bob managed to get it and became well known as a glacier pilot. There were mines that were on the edge of glaciers, and Bob was able to supply them year-round by using skis on his plane, flying off the mud flats when the tide was out. As I mentioned earlier, Bob did a lot of flying for the government during the war when they were building airfields around the interior and in the Aleutian Islands and later set up Reeve Aleutian Airways, which was quite successful.

I did not do any flying after I returned to Fairbanks as it was expensive to rent a plane, and by the time I could afford to do so in New York ten years later the air traffic was such that I did not think a Sunday pilot was very safe. One had to fly regularly to keep up with instrument flying and I did not have the time to do so. However, the company had several Aero Commander aircraft over the years and I made frequent flights to South America as copilot in them. I was checked out in them and so was able to take over in an emergency. I only had to do so once, when the regular pilot and I were flying from Colombia to Miami. We had had lunch in Jamaica, and just as we were entering the traffic pattern at Miami the pilot became very sick. The seat behind the copilot was a toilet, so the pilot was busy on it and also giving me some guidance, so I was able to let down and land the plane safely. One other time in Mexico I was in the copilot seat in a new twin-engine Cessna with Cappy, as we called the pilot. We were flying into Puerto Vallarta, and about 40 minutes out Cappy became very ill and was filling all the airsick bags we had. He could hardly hold his head up, and I flew until we were on our final approach when Cappy rallied, saying, "As sick as I am I know that you have not been checked out in this plane so I will land it," which was fine with me.

South American Placers, the International Mining Bolivian subsidiary of which I was president, bought two tri-motored Northrop aircraft. They had been designed and built for troop and vehicle transport in Korea, but the war had ended before the planes got into service. They were never licensed in the United States but we got permission to fly them to Bolivia, where they were licensed. They made many trips from the end of road transport at Caranavi to Teoponte, moving over 6,000 tons of dredging equipment. I was checked out in those planes as I flew in them frequently, but they were more work than pleasure to fly. As much as I liked piloting aircraft, when I made up my mind to go back to mining the decision was final, and I never looked back or had any regrets.

However, one time when knowing how to fly was very helpful was in Africa in 1982. I went over with my family to visit some friends who were living in Johannesburg. They had arranged safaris in South Africa, which were great, and then they arranged for a pilot to take us to Botswana; they were to join us the following day in another plane. We went out to the airport early in the morning and were greeted by a very young man who told us that the pilot who was supposed to take us was ill and that he had been asked to take us. He looked very young, and we were quite skeptical, but felt sure that our friends would not arrange for our family to go with someone who was not qualified. He told us that he had flown to the camps we were going to, so we set off for Botswana. We were to stop at Maun and then go on to Savuti, a game camp about 30 to 40 minutes flying time further, where we were to see game and spend the night. We arrived in Maun without incident, and after getting gas we took off for Savuti. I was in the copilot seat, and Sandra, Erin, and Kevin were in the back. We expected to be at the camp well before lunch, but after an hour the pilot said that he had underestimated the time but that we would be at the camp soon. This went on for almost another hour, by which time the pilot had changed course several times and was getting very nervous. At this point I realized that he was lost and that we were out in the middle of the Kalahari Desert with two children with diabetes who had to eat. When I questioned him, he finally admitted that he had never flown to Savuti. I really got mad and cussed him out thoroughly, and he fell apart and admitted that he was hopelessly lost, and then said that his direction finder had stopped operating and he did not know how to get back to Maun. He had told me when we started out that there were three hills close to Savuti, but we did not see any hills in the time we were supposed to and he had kept on flying towards and around three other hills, which, if they were the ones I thought they were on the map, were way off to the northeast of where Savuti was supposed to be. I had been watching the direction we had been flying and following the map as well as I could, which was not easy as it was terribly dry and the streams shown on the map were completely dry. The pilot was so upset and confused that he had no clue as to where we were, so I took over and set a course for Maun, assuming that he had been off course from the start. We had been gone over two hours and supposedly had three and a half hours of fuel, so I knew we had to get to Maun or at least closer to where we were supposed to be so that we would have a chance of being found if we ran out of gas and had to land in the desert. Sandra always made sure that we had food and water with us wherever we went, but with two children with diabetes I knew we could not get by very long in the desert so I was very concerned. I could not get any help from the pilot, but fortunately in less than an hour I saw landmarks I had seen on the way out, and we got back to Maun with about 15 minutes of fuel left.

I was on the board of a shipping company when I was with Rosario in the early 1980s and went to Japan for the launching of a ship that the company had built there. Many government and business people were there for the launching and the big party afterwards. I was talking with one man about flying, and he said that he had been in the kamikaze group and was scheduled to make his one-way flight the day the atomic bomb was dropped (his flight had been called off). I said that I could not understand how the pilots could have been indoctrinated so that they were willing to make a one-way, sure-death flight. He said that they had been told how the Japanese mistreated, raped, and killed many people in the countries they had occupied and that if their country were occupied their families would be subject to the same horrors. Japan was not doing well in the battles at that time and the loss of the war was a real possibility, so he said that he had been ready to go but was certainly happy when his flight was canceled. He also said that the Japanese were very grateful and appreciative of the way they were treated by the United States after the war.

The head of the agency that handled the shipping company's business in Japan visited Sandra and me in the hotel when we arrived in Tokyo to greet us and give us our train and plane tickets to go to the launching. After he left, Sandra said that she would like to know what kind of cologne he had on and asked if I would please ask him while we are together the next few days. A day or two later I told him that Sandra had really liked the cologne he used and wanted to get some for me. He said that he never used cologne but that he had been with his geisha just before meeting us at the hotel, so obviously I did not get that sort of cologne.

4 Mining

ALL of us seven boys as well as our oldest sister worked in Dad's store in Cordova from the time we could stock shelves or count items at inventory time. Later we moved to clerking and so forth. We all worked after school every day, on all Saturdays, and often on Sundays and full time during summer vacation. Dad hoped that all of us would stay on with him in the store, but I did not like working there. I especially disliked waiting on people. Once a week a steamer would arrive in Cordova from Seattle with supplies for the town, and we would receive fresh fruits and vegetables, among other things. After a week on the ship there would be considerable spoilage, so we would clean and trim everything before displaying it. After cleaning the fruit I would often arrange it in nice pyramids, only to see the ladies come in and pick through most of the pile, dropping many apples or oranges on the floor for me to pick up and re-pile. I would have to discard the ones bruised too badly and I often felt like growling, but Dad always said to keep on smiling at the customers. I usually tried to work in the warehouse or at cleaning sidewalks and the like. One day in the spring when I was 15, I was shoveling snow (over two feet of it) alongside the store when Dad was going by with a man that I knew to be Mr. Cramer, superintendent of a small gold placer mine in the McCarthy district where my older brother Bill was a foreman. As they were passing Mr. Cramer said, "Gee, Harry, that kid is very handy with a shovel; is he one of yours?" Dad said, "Yes, this is Pat," and he introduced us. I asked Mr. Cramer for a job in his mine, and when he asked how old I was I said 18. My Dad blinked, but with 12 kids he wasn't sure how old anyone was and did not say anything, although I am sure that he suspected that I was not 18. Mr. Cramer said that he could use me if it was all right with Dad, who said it was OK with him. So I got my first job in mining at the age of 15 and have been active in mining ever since.

The Chititu Creek mine was a small open-cut gold placer mine about 20 miles from McCarthy. It was quite isolated, the last few miles being

101

passable only by foot or with a horse-drawn wagon. Bill wasn't very happy when I showed up. He didn't like nepotism and thought I was too young. He said that if I didn't work harder than anyone else on the crew I would be fired, so I worked as hard as I could and have been doing so ever since. It was a good start for my mining career but a tough one at first, as all the men on the crew were older and more experienced. However, I held my own and soon was able to do my share and more. We worked ten hours a day, seven days a week, and lived in tents; I shared one with five men. There wasn't anything to do except work and sleep, and nothing to buy except gloves, socks, and Copenhagen snuff. As foreman Bill had a separate tent but wasn't very receptive when I visited him. I was lonesome and wondered if my parents missed me as much as I missed them. We got word that two men had been killed in a similar mine in the next valley over from us, and I was sure that my parents would be concerned and might even consider calling me home when they heard about the accident. About a week later a man came walking into camp just as we finished lunch. He asked for me, and I thought that my parents had sent for me, but when I stepped forward he said that he was an insurance salesman and that when Dad had heard about the accident he had thought that I should have some insurance. I got the message and quickly got over my lonesomeness. One very large man called Big John thought that everyone should use snuff, and he would pick me up with one hand and put snuff in my mouth with the other. It made me sick the first few times, but then I decided that I had best learn to use it, which I did for several years. It was a harder habit to break than smoking, which I quit years later.

One of the fellows I worked with during the four seasons I was at Chititu was Jorge Krohn. He was an experienced logger and a hard man to keep up with, but we got along well and would occasionally go for a walk after dinner. One evening we found some wild raspberries, which were really tasty and the first fresh fruit we had had for a long time. We mentioned the berries to the cook and he said that if we could find some ice he would make ice cream, if we thought we could find enough raspberries to have a treat for the whole crew of 14. The next few evenings we went exploring up Rex Creek, as we had been told that years earlier prospect adits had been dug into the hills alongside the creek. We found one that had ice in it, so the next evening we picked buckets of raspberries and the night after that took sacks to get the ice, and the following night everyone had the treat of the season with fresh raspberries on ice cream. The entire crew was most appreciative and it tasted great to me.

The men I was working with talked about spending their winters in the southern areas of the country. They could save up to 800 dollars during the summer, which at that time was enough to spend the winter in a warm climate taking it easy. It sounded good to me, and I was thinking about doing the same, but one evening one of the older men told me that it had been fine when he was young but that the older he became the harder it was to labor ten hours a day, seven days a week. His advice to me was to get an education, and that is when I decided that somehow or other I would go to college. I have always been appreciative of his advice.

The day I was leaving to return to college after my second season I packed up my belongings and then went to pick up my pay and say good-bye to everyone before heading down the creek. I had several miles to walk to the mouth of the creek, where a car was to pick me up. I thought that my bag was heavier than usual, but it wasn't until I got home and unpacked that I found an eight-foot piece of heavy chain weighing about 30 pounds carefully wrapped in my clothes. I then realized why the fellows had all been smiling when I left.

After four seasons at Chititu I worked in the Fairbanks area. Jorge stayed on at Chititu, and one day when they were blasting very large rocks one of the dynamite charges went off prematurely, and Jorge was badly injured. He lost one eye and ended up with a glass eye and many scars. Several years later, after Chititu had closed down, Jorge was working on a crew with the U.S. Smelting, Refining, and Mining Company at the time that I was dredge engineer. We were moving a dredge from one valley to another during the winter. We would go out to work if the temperature was above 40 degrees below zero. Usually the temperature would rise a bit during the day and occasionally it would fall, but we would always finish the shift even though it would sometimes drop to 50 below or lower. Jorge wore a wool hat with earflaps, and when it was really cold he would twist his hat so that one of the flaps would be over the glass eye. He said that the glass eye would get colder than his ears, but it always looked strange and he sometimes had frosted ears.

Glass eyes remind me of a hard-hearted banker with a glass eye that I knew for many years in Fairbanks; I later conducted his burial service when I was president of the Pioneers. A man that I knew tried several times to get a loan from this banker but was always turned down. He was complaining about the banker being too hard-hearted when the banker said that one of his eyes was glass and that hardly anyone could tell which it was when they looked at him straight on. He told the applicant that if he could look at him straight on and tell him which was the glass eye he would give him the loan. The man looked and said that the left one was glass, and the banker asked how he could tell. The man said that he could detect a little sympathy in that eye and there wasn't any in the other. He got the loan.

My first summer job in the Fairbanks area was taking a man from Boston on a prospecting trip. George Pond had been too young to participate in the Klondike gold rush of 1897, which he regretted. He was a World War I veteran who had done well financially and wanted to satisfy his desire to go prospecting, so he contacted the University of Alaska to find someone who would take him. Dean Patty asked me and one of the short-course miners, Carl Tweiten, to take him prospecting. Carl had been prospecting and trapping in the Goodpaster River area and had a poling boat at Big Delta, where the highway crossed the Tanana River just below the mouth of the Goodpaster River. We therefore decided to take Mr. Pond up the Goodpaster River to prospect in that area. We assembled the food and equipment and set out with Mr. Pond. We engaged a man with a powerboat to take us from Big Delta with our supplies and tow the poling boat as far up the Goodpaster River as possible. From there we poled upriver for several days. We would pitch a tent at night on a gravel bar, cut spruce boughs to sleep on, and set up a small iron stove to cook on. The mosquitoes were unbelievable. We had to wear head nets and gloves and put up mosquito netting to sleep under. The only repellant against mosquitoes at that time was Buhack powder, which was quite effective when burned in an enclosed area, so with the Buhack and our bed nets we were fairly comfortable sleeping in the tent. The first morning out Mr. Pond said that he would like pancakes for breakfast, so I prepared the batter and started cooking them on top of the stove outside the tent. By the time the pancakes were ready to flip the uncooked side was covered with mosquitoes. I tried to scrape them off but quickly saw that if I kept doing so there wouldn't be enough of the pancakes left, so after one or two attempts I went ahead and flipped the pancakes, mosquitoes and all. Mr. Pond said that the pancakes were very good but asked what the black specks were, and I told him that I had put in some chopped raisins. He soon caught on to what I was doing and was somewhat annoyed, but finally said that the pancakes were very good and, as he suffered no discomfort from them, would I please keep on making them. He named them "Pat's mosquito pancakes." After he returned to Boston at the end of our trip, he gave a talk on his experiences prospecting in Alaska and mentioned my mosquito pancakes as a memorable part of his trip. It was hard and slow work poling the boat up the river. When we encountered rough water or rapids, one of us would pull with a rope from the shore while the other poled and Mr. Pond walked along with the one on the shore. When we had to cross small creeks, one of us would have to carry Mr. Pond on our back, as he had nice leather hiking boots that he did not want to get wet.

We panned for gold along the river and in the streams that entered from either side. The first time we found any specks of gold was in a stream that came in on the right side looking downstream. There was an old trapper's cabin there, so we fixed it up as well as we could, using cheesecloth for the windows to keep out the mosquitoes. We fixed up three bunks with willow poles to hang our individual mosquito nets, and with the Buhack powder we were quite comfortable. For several days we dug and panned for gold up the stream we were on and continued to get enough indications that we wanted to move on upstream to see if we could find any veins that might be the source of the gold we were finding in the stream gravels. Mr. Pond did not want to go backpacking with us, however, so we fixed him up with food and the Buhack powder and Carl and I set out. We worked our way up to the headwaters of the creek, prospecting as we went. We would camp close to the creek for water. Often, though, we would climb up on a knoll or ridge to sleep where there was a breeze and the mosquitoes weren't so bad. We carried along a piece of canvas that we used to make a lean-to for protection from the rain. We passed a few pleasant evenings with a fire in front of the lean-to, and we watched the late sunsets and the early sunrises not far apart in time as it would not get very dark in June in that part of Alaska. I carried a small volume of poems by Robert Service and would memorize and recite poetry while we enjoyed the late evenings. We finally found an outcropping of a vein, and took samples and staked some mining claims in the names of the three of us. However, when we returned to the cabin after ten days or so we found Mr. Pond rather distraught, because the Buhack was all gone. We could not understand how he had used it all up since we had left three cans, and it did not take much to keep the mosquitoes under control in the cabin. He finally told us that the mosquitoes were really annoying him when he took down his trousers out in the woods, so he would light a pile of Buhack to squat over and he had used it all up. He added that he had really enjoyed his adventure but wanted to return home, so we had a very pleasant two days floating down the river to Big Delta. There were wild roses and many other flowers along the banks of the river. We were moving so silently that we came on two bears, a moose, two foxes, a beaver, rabbits, a muskrat, and weasels. We stayed at the roadhouse at Big Delta known as Rika Wallen's for three days while waiting for transport to Fairbanks. Rika was from Finland and had been running the roadhouse for many years. Many years later I stopped by there with my family, and they were impressed that I had stayed there almost 50 years before as it had become a national historic site maintained by the Forest Service. I learned a lot from Carl about many things on our trip—how to live, work, and travel in the rivers and hills—but the most important thing I learned from him was to always try to do more than my share. When both partners are trying to do more than half, the work gets done in no time at all. We would get the tent set up, the spruce boughs laid, the stove set up, wood cut, and dinner under way in no time at all. It was a great lesson that has been helpful all my life. Mr. Pond said that he would probably never return to Alaska, so Carl and I filed the mining claims in our names. Being in college I did not have the time or money to do the assessment work, so I later transferred my interest to Carl. A mine was later opened up close to that area, and Carl was working there when he lost most of his vision in an explosion; he subsequently gave up mining and ended up in real estate near Seattle, where he could manage with very limited vision. As of this writing in 2006 a substantial mine has been put into operation in that area.

When we arrived in Fairbanks after the prospecting trip I started looking for a job for the rest of the season. It was during the Depression, and there were lines of men every day at the Fairbanks Exploration Company (F.E. Co.), and it was very difficult to find a job. Fortunately I ran into the employment manager on the street one day; he was also the manager of the Elks basketball team and I knew him as I was athletic manager of the college team. He told me that as soon as there was an opening he would put me on. While I was waiting I stayed at a small hotel on the banks of the Chena River, and one day Wiley Post and Will Rogers arrived and landed on the Chena River near the hotel. Joe Crosson, manager of Pacific Alaska Airways (a subsidiary of Pan American Airways), was looking after service for Wiley Post. I had serviced Crosson's plane in Cordova when I was working for Harold Gillam, so I went down and told Joe that I wasn't doing anything and would be glad to help. So I helped gas up the plane the morning they were leaving. They could not get off with full tanks because there wasn't a long enough straight stretch in the river. We pumped out a major part of the gas and drove it out to Harding Lake about 20 miles out of town and then put the gas back in the plane after they had flown out. They took off on their way to Point Barrow on the northern coast of Alaska. There weren't any navigation aids in those days (1935), and Wiley landed on a small lake not far from Barrow to ask directions from some natives at the lake. When he took off, he banked very sharply, and as the plane was low on gas, the engine coughed and the plane crashed. They were both killed. They were great men, well known, and well liked and respected all over the world. It was a very sad day.

Charlie Fowler called me a day or two later and said that there were openings for two men on the bull gang at Fox and that he was sending me and another man out. He told us that it was going to be very difficult as the entire crew including the foreman was Finnish, and they wanted an all-Finnish crew and would do anything they could to get rid of anyone who wasn't Finnish. They worked very hard, spoke only Finnish, and tried to make things as difficult as possible, so it was hard to keep up with them let alone do one's full share. I had been doing similar work at Chititu and knew what had to be done, however, and I gave them a run for their money. The fellow that had gone out with me quit after a few days and another Finn replaced him. They just about wore me out with their ten-hours-a-day, seven-days-a-week work shifts, but I needed to work so I could go back to college in September. After another week they said that they had accepted me and would go back to a more normal work pace and speak English, so I finished out the season with them.

The winter that I had to work after the hockey trip I worked at an underground mine as a mucker, then as a miner for awhile, then as a millman, and later as the engineer assayer. It was all excellent experience. The millman that regularly worked the night shift once asked me to go to Fairbanks with him. He said that his girlfriend was an excellent cook and had told him that he could bring a friend for a chicken dinner. He did not tell me where she lived until I questioned him as to why we were heading to the "line," as it was called, a fenced-in row of cabins for prostitutes. He said that his girlfriend was one of the girls and that I could back out if I wanted but would be missing an exceptional dinner. I was under no obligation to stay after dinner, he said, as he was spending the night with her. He later married her, and they were getting along fine the last time I ever heard of them. Incidentally she really was a great cook.

The "line" traced its origins to the very beginnings of the town of Fairbanks and had its equivalents in all the frontier towns of Alaska and the Yukon. Most citizens accepted the presence of a restricted district for prostitution. As early as 1914 there were efforts to close the line in Fairbanks, but public sentiment favored the continuation as a necessary evil for the safety of the very few women living amongst thousands of men. However, rules were established regulating the life of the prostitutes, requiring them to have regular blood tests and pay a vagrancy fine of 50 dollars every month at the police station. At one time after many complaints the authorities rounded up 23 prostitutes and charged them with operating bawdy houses. Women on the jury said that their daughters would not be safe if the line was closed, and the defense attorney quoted Jesus, who said that he who is without sin should cast the first stone. There was not a single conviction in 23 cases because witnesses were not willing to admit personal knowledge of the line. The commanding officers of the nearby army bases complained and threatened to declare Fairbanks off-limits at various times, but as Fairbanks expanded and the pressure on the line increased more bars and houses of ill repute opened up outside the city limits. By the mid 1960s the remaining cabins were cleared in an urban-renewal project. I knew dance-hall girls in Dawson as well as ladies of the night in Fairbanks who married and became very respectable women and socially acceptable in their communities.

One spring the man in charge of exploration told me that the only opening he had for a drill helper was on a crew that he was sending to the

Kuskokwim area; the helper had to be the cook as well. I really needed a job, so I told him that I could cook and he hired me. I went to the bookstore, bought a cookbook, and got help from the man responsible for all the camps. He worked with me in ordering all the supplies and fixing up a Yukon stove with a steel plate on top for frying. A Yukon stove is about a foot square and two and a half feet long, with the firebox in the front half and the oven in the rear half. The heat from the firebox goes over the top of the oven and it is difficult to control the heat for baking, especially when you are inexperienced to start with. Our daily routine was that after breakfast I would pack lunches for the crew, four including me, and then go to the drill to work until an hour before quitting time to go back to camp to fix dinner. One day a week I would stay in camp to bake, but if they needed me at the drill they would ring a bell and I would leave my baking and go to the drill. Baking bread in the little stove was difficult, and it took me several tries before the bread would peek up over the side of the pan. I also baked pies, cakes, and cookies. During the time I was learning how to make a decent-sized loaf of bread I did not have enough bread for sandwiches, so I would make extra pancakes for breakfast and make sandwiches out of the cold pancakes. The crew soon started complaining, but fortunately by then I was having more success with the bread. One day when I was baking I saw a recipe for brown Betty, I believe it was called, a single-layer chocolate sort of cake. When it was about half baked, the bell rang and I had to go to the drill. By the time I returned the fire was long out and the brown Betty had dropped. I left it in the oven and lit the fire for cooking dinner. A little while later I took it out and put it on a cutting board. It was so flat that I thought it was useless, but that evening one of the fellows asked me what it was and could he try it. I said, "Go ahead," but he could not cut it with a knife, so he chopped off a piece with a hatchet and chewed on it for quite awhile. It turned out to be the best thing I made all summer as one small piece each evening kept them happy. It was one of a kind, though; I tried to duplicate it without success.

We were prospecting on claims owned by Cecil Barlow, and to start with we were camped close to his cabin and visited with him often in the evenings. He was from England and had a large collection of English literature and poetry. He had memorized many poems, which he would often recite to us. Whenever he went to the Pioneer meetings in Anchorage, they would always call on him to recite. Cecil and his partner had mined in a small way for many years. One winter day when Cecil awoke, his partner was dead. Men who worked and lived together in isolated areas had been known to kill one another, and Cecil was very afraid that he might be accused, so he took the body out behind the cabin and wrapped it in canvas to protect it from animals and, hopefully, to freeze, until someone could verify that there hadn't been any foul play. Cecil's cabin was not far from the trail between Flat and McGrath that the district judge and his retinue would travel along when holding court in different places. The judge was Charles Bunnell, who later was the founding president of the University of Alaska. Cecil could see the trail on a ridge several miles away, and when he saw the dog teams he went up to the trail to intercept them. He told the judge what had happened and pleaded with him to go and look at his dead partner. The judge said that he did not have time as the daylight hours were very short and they had to move on, but they all knew that Cecil was a very honest man and believed his account of the death, so the judge ordered his clerk to fill out a form certifying that it was a natural death. When Cecil asked the judge what he should do with the body, the judge said that he knew it would be difficult to dig a grave in the frozen ground, but since the man was well frozen he could just sharpen him on one end and drive him in. Cecil ignored the preposterous suggestion and thawed the ground with wood fires, just as he did in sinking prospect shafts, and gave his partner a decent burial.

We were drilling about five miles from a small mining operation near an airstrip. One of my chores was to carry gasoline for our drill in a backpack from that operation, as they hauled in their supplies by tractor during the winter. One day a piece of the drill mechanism weighing about 60 pounds broke, and I took it to the mining operation to get it welded, expecting to be back at our camp by dinnertime. However, one of their own units had broken down, and they had to repair it before they could help me, so I had to wait and stay overnight. During dinner that evening someone said that one had to be very careful walking through the brush as it was rutting season for moose, and they were very dangerous if one ran into them unexpectedly. As soon as the part was repaired the next morning I headed back to camp, and each time I had to go on a trail through brushy areas I would stop and listen for any noise of a moose. At one place I heard a "clop clop" sound in a swampy, brushy area and immediately climbed a tree, pack and all. When I looked down, there were the fellows from my crew who were out looking for me. They had a good laugh, and I was embarrassed but at least safe.

When I was too young to work in my father's store, I worked on Saturdays for Joe Frye and Nick Stitch, two men close to our home who raised chickens. In the morning they would kill the chickens and dip them in boiling water and I would pluck them; then they would clean them and in the afternoon I would deliver the chickens to the customers on foot, as one could walk anywhere in Cordova in 30 to 40 minutes. One of the regular customers was a woman on "the line" named Billie who was the most generous of any of the clients with tips, and, of course, she always got prompt delivery. As I got bigger and was working in Dad's store, I often delivered groceries to Billie. The business district of Cordova was only three blocks long and on the western end was "the line," which was a row of small cabins separated from the main street of town by a six-foot solid fence. The ladies of the night, as some called them, plied their trade out of sight of the main activities of the town. Whenever I had some spare time, I would break up wooden boxes and make kindling, so when I took an order to Billie I would take along a small bundle of kindling. The best tips I ever received were from her.

When I was 15, I started to work during the summers in a small gold mine in the McCarthy district, north of Cordova near Kennicott. The second year when I was on my way to the mine, a railroad bridge across the Nenana River at Chitina had washed out, and we had to go across in small rowboats. As I was getting off the train I saw Billie, who was asking for help getting her baggage and herself to the riverbank and across the river. I am certain that most of the men, who ignored her, knew her much better than I did but were afraid of gossip that might get back to their wives. I carried Billie's bags to the boat, steadied her in the rough crossing, and helped her to the connecting train. After the fishermen left town for the summer fishing season, some of the women from "the line" would go to the mining camps in the interior for the summer, and I assumed that that was what Billie was doing.

Later on when I was 20 I was working on a prospecting crew on Moore Creek in the Kuskokwim region. I was the cook and helper on a prospect drill crew. We moved across country from Moore Creek prospecting on several streams. We had an arrangement with Star Airlines to drop us meat and fresh vegetables on their weekly run from Anchorage via McGrath to Flat. One week the weather was very bad, and the plane did not show up. Two or three days later, just as we were finishing dinner, another Star Airlines Bellanca circled low over our tent and dropped a note in a weighted bag telling us that the pilot, Dan Victor, had turned up the wrong valley in bad weather and crashed in the headwaters of the creek on the other side of the ridge to the east of us. The note asked if we would rendezvous early the next morning at the top of the ridge to the east of us with men from the mining operation on Moore Creek, and he would fly over to lead us to the crash site. We waved our agreement, and I packed some food in a backpack, and the four of us in our crew started out for the ridge. We walked all night and met up with five men on the ridge about 8 a.m. The pilot flew over shortly after and indicated the direction to the crash site, so we walked down into the valley and about three hours later found the wreck. The pilot was walking around with his arm in a sling, and Billie, who had been thrown through the windshield and was badly cut up on her face, had a broken leg and was propped against a tree. When she saw me she exclaimed loudly, "Here's Pat O'Neill, who saved my life once before and now has come to save me again!" The other passenger was still in the wrecked plane as he had been killed by the cargo when the plane had crashed. The pilot who had led us to the site dropped a roll of canvas so that we could make a stretcher, a shovel to bury the dead man, and some food. While the man was being buried, another man and I cleaned up Billie as well as we could. She had one major slash from above her left eye across her nose and on to her right jaw, as well as many smaller cuts and bits of Plexiglass imbedded in her face and head. We cleaned her up as well as we could, made some splints and tied them on her leg, and put her on the improvised stretcher, which was just a piece of canvas tied on to two poles cut from small trees. As we started out, she insisted that I be on one side or the other close to her head. She was not a tall woman but was fairly heavy and was a load for four of us to carry, especially going uphill. It was early afternoon when we started climbing up the hill, and we walked until we reached the ridge late in the evening. We stopped to rest and eat what food we had. Whenever we stopped, Billie would ask me to pick out glass from her face and head. At that time of year, August, it was dark for only two or three hours, so we rested until there was enough daylight to walk safely, struggling along with the stretcher. We arrived in early afternoon at the airstrip on Moore Creek, where the plane was waiting to take Billie and Dan Victor to the hospital in Anchorage. According to my parents, who had moved from Cordova to Anchorage earlier that year, Billie called them from the hospital to tell them that I had helped save her a second time. She told them about her crash and rescue, and they visited her in the hospital.

I didn't hear anything about Billie until four years later. I was working at a gold mine near Fairbanks and would go to town fairly often. I had a date with a girl who was working at the Chena Bar, where I also worked occasionally during college when I needed money. One evening as I was sitting at the end of the bar near the door waiting for my date to get off work, I looked back over my shoulder to see what the commotion was at the door, and there was a very drunk Billie with two men who were just as drunk. Her face was badly scarred, and she had gained weight since I had seen her: she was bulging out of a sweater and britches. She was quite a sight and as soon as I spotted her I pulled up my coat collar and leaned over the bar hoping to avoid her, but she had seen me and grabbed me in a bear hug and proclaimed loudly that here was the most wonderful man whom she had known for many years and who had saved her life twice, and so on. By the time I broke loose, my girlfriend was gone, and that was the end of that romance.

The only thing I heard about Billie for several years after the bar incident was that she was drinking even more heavily and had moved in with a fellow named Tex, who also was a heavy drinker. Then I heard that Billie and Tex had won the annual ice pool, which is an organized gamble where people bet on the day, hour, and minute that the ice breaks up and starts moving down the Tanana River at Nenana about 50 miles south of Fairbanks. It was the biggest pool up to that time, about $200,000 dollars, and Billie and Tex were the sole winners. People who knew them were certain that the money would be wasted in a grand drunken orgy, but they surprised everyone and called my parents to ask if they would go with them to the church to witness their wedding and their pledge of non-drinking. My folks gladly went along, and Billie told them to be sure and let me know, as I had been a positive influence in her life. Billie and Tex then moved to southern California where they bought a small house and lived for many years soberly and happily, according to news I heard occasionally.

One other time, when the plane that dropped us meat did not show up because of bad weather, I killed a bear so that we would have some meat. I did not know much about butchering a bear but cut off a piece that I thought would be good and grilled it. It was so tough you could hardly chew it, so I cut some up in chunks and tried stewing it. I would keep it cooking on the stove, even getting up at night to keep the fire going, and then I would take the pot to the drill during the day and keep a fire going under it there. The fellows would chew on it without much success at dinner. One evening a bear cub wandered into camp, and it was much better eating, but everyone was happy when our weekly drop of meat and vegetables resumed.

Admittedly, I was not a great chef by any means, but the only one to complain was the foreman of our crew. He would often say that his mother would fix things this way or that way; I would try to do the same, but I was getting increasingly aggravated by his complaints. One day when it was getting close to time for me to head back to college, I was serving spinach and he said, "Now my mother would have prepared the spinach in such and such a way." So I said, "You need to send for your mother because I am leaving." I hit the trail early the next morning with my duffel bag en route to Flat to catch a plane to Anchorage for a day with my parents, and then went on to Fairbanks and janitoring and studying for another year. The next summer season I worked as a panner on prospect drills on various creeks in the Fairbanks area. The churn drills would extract the gravel two feet at a time from the holes being drilled, and the panner would pan the material to recover the gold so that the value of the ground could be calculated. It was rather tedious work, so I kept a book of Robert Service's poems on a shelf in front of me and would memorize poetry while I worked. I have had much pleasure in reciting poetry on various occasions ever since.

The following season I went on various exploration trips looking at potential mining properties. One trip was on a tributary of the lower Yukon River near Aniak with a very interesting man, Don Draper, who had many years of experience in exploration. He kept very detailed notes about the flora and fauna, the weather, the character and the flows of the streams we crossed, and everything of any consequence at all, as well as the type and character of the gravels we were sampling and the gold we found. His reports were the most complete I have ever seen, and he was a great fellow to travel with. There were many grizzly bears, and we carried high-powered rifles. The lead man carried his rifle in his arms ready to fire as you never knew when you would encounter a bear, especially where there was much brush in hilly areas. We saw a few bears at a distance as we were hiking over a three-week period but did not run into any until one evening. We were camping on a river bar, and after dinner we were doing some panning. Our guns were at our campsite, and we did not realize how far we had gradually moved away until we suddenly heard a noise, and about 20 feet away three big grizzly bears were growling and standing up on their hind legs. They must have been eight feet tall and were very threatening. We thought our time was up, but we ran back to our guns faster than either of us had ever run before or since. When we reached our guns and turned to look back, the bears were crossing the river. We were very fortunate and were never without our guns again that trip. At night we would take turns standing guard, as we were sleeping out in the open. Don Draper spoke very little about his early life; he was from the eastern United States, was well educated, and had been an active member of the Explorer's Club in New York. When I met him he was living in a small one-room cabin on the banks of the Chatanika River, about 40 miles from Fairbanks. He loved fishing, and I used to go out to visit him and catch some nice trout or grayling. He did a few exploration jobs for the F.E. Co. in addition to the one he and I did, and he also did some research for the university on animals and occasionally sold wildlife pictures to some magazines. He was fairly tall and could not stand upright in his cabin, but he lived in it for 50 years that I know of and was quite a recluse. He died in Fairbanks several years ago.

When I was working at the McCarty mine, the winter I had to stay out of college paying debts from the hockey trip, there were eight of us in a bunkhouse. Four of us obtained a chess set and learned to play. After two months or so we had a tournament, and near the end of it, one of the miners who kept a gallon of whiskey under his bunk and drank heavily every night staggered by and said that he would like to challenge the winner. We were amused but agreed. I happened to be the winner, so Einar and I played the next evening. He beat me so quickly I wasn't sure how it happened. I spent time afterwards with him learning more about chess and also about him. He had been a university professor in Denmark but had gone through a bitter divorce and become an alcoholic and a drifter and ended up in Alaska where he became a miner. A few years later after the war when I was dredge superintendent, Einar was working on the dredges. He was very capable and could do almost anything but would take off on a drunken spree from time to time and so would be fired, but later, when I needed a good man, I would search him out and hire him back. One afternoon I was taking the general manager on a tour of the dredges; we were leaving Dredge 6 just as the 3:00 p.m. shift was coming aboard, and among the crew was Einar, very drunk. I had no choice but to fire him. He said, "Pat, you have fired me several times, but I always came back when you needed me. I am warning you that if you fire me now, I am never going to give you another chance." And he never did. I tried two or three times to hire him again, but he would never accept, saying that he told me he would not give me another chance. His drunken periods increased, and he burned to death in a cabin fire a few years later.

Also when I was working at the McCarty mine we had a party for one of the mill men who was getting married to a very fat woman. The hoist man commented about the size of the bride-to-be, and Neil said, "Yes, she is very big, but when you fellows are shivering in your sleeping bags in the cold weather I will be cuddled up to my bride and be nice and warm." The hoist man, who had a high, squeaky voice, then said, "Yeah, Neil, but you will not know if there is someone on the other side of her." At their wedding the bride, who always wore big, black shoes that looked like men's shoes, had on high-heeled shoes and was delicately walking down the aisle in very obvious pain. As soon as the minister pronounced them married, and they turned to face the audience, she kicked off her high-heeled shoes and hollered to her younger sister to bring her some other shoes, which she put on before walking out.

During the winter I operated the mill at the McCarty mine. I worked the 3:00 to 11:00 shift for a few months, and occasionally after my shift when there was a bright moon I would ski several miles down the valley to visit with a friend who was operating the mill at another mine on the midnight shift, 11:00 to 7:00. It was pretty with the moon on the snow. I enjoyed the exercise and the visit over a cup of coffee with my friend, but I would get somewhat worried skiing at night when wolves were howling. The winter

I was working at the McCarty mine we had what seemed like a very long spell of very cold weather, and I specifically recall, without any pleasure, having to make many trips to the outhouse one night when I was ill and it was 50 below. A year or two later I stayed overnight at a cabin on Big Eldorado Creek when it was very cold, but the owner had lined the toilet seat with wolverine fur, which does not frost up, so it was a more comfortable experience.

While I was working at the McCarty mine the first radio station in interior Alaska, KFAR, came on the air, and what a wonderful joy it was to have news and music in the isolated mining camps. The first song I heard on KFAR was Judy Garland singing *Over the Rainbow*, and I will never forget it.

Not very long after I was appointed dredge superintendent the president of the company in Boston visited the operations, and I was assigned to take him to all the dredges one day. It was a grueling day as he asked many questions about the operations that I did not know the answers to as well as I should have, but I struggled through the day. I was relieved to drop him off late in the day at the general manager's house and go on to where I was living, about a block away. I had just fixed a martini and was taking off my boots when the president knocked at my door. My first thought was to wonder what more questions he wanted to ask, but he said, "I just found out that the general manager and his wife not only do not drink but do not have any liquor in their house. I told them I remembered something I wanted to discuss with you. You look like a fellow who would enjoy a martini, and after my tough day with you, I need a good martini." What a relief! We had a relaxing time over several martinis.

I was 33 when I was promoted to dredge superintendent. Several people commented that I was the youngest superintendent ever named at the company, and I was feeling quite jubilant. One older man told me that the reason I got ahead was because I worked harder than anyone else. Then I went in to see the manager, who was retiring. I told him that I sincerely appreciated the confidence that he had expressed in naming me to be a superintendent. He said, "Don't let it go to your head, young man; you just happened to be in the right place at the right time. Personally, I don't think a man is worth a damn until he is over 40, and you have a way to go." He brought me down to earth in a hurry. Shortly afterwards I was invited to

Gold dredge in Alaska.

the manager's house for lunch. I was a bit nervous, it being my first time out in the company high society, but everything went well until we went into the living room for coffee. One of the ladies commented on the many lovely items of antique furniture that Mrs. Ogden had. She thanked the lady and spoke about some of the items, and added that her most prized item was the nice little chair that I was sitting on. I twisted a bit to look down at the lower part of the chair, and it collapsed, so I was suddenly sitting on a pile of damaged wood on the floor. It was the most embarrassing moment of my life.

The U.S. Navy had a group based in Fairbanks in connection with the development of oil in the naval petroleum reserve on the North Slope. I had met the captain in command, and he asked if I would take him out to see the dredges in the wintertime so that he could take some pictures. One nice day, therefore, I called him and he asked me to pick him up at his house as he was going to put on some warm clothes. When I entered his house, his wife Alice was sitting down sewing. She asked if I would like a cup of coffee, and when I said yes she stood up and plunged the needle into her left breast. I gasped, and she said that she had not meant to startle me but that her falsies worked well as pincushions. I mentioned this incident to Sandra after we were married; many years later we ran into Alice and Ralph in a little hotel in Copenhagen, and when I introduced them Sandra had a hard time keeping a straight face.

Before the F.E. Co. started mining on a large scale with dredges, mining had been done by sinking shafts through the frozen material to bedrock, drifting out in four directions, and digging out the ground retreating towards the shaft. The material was hoisted to the surface and dumped in a pile over sluice boxes for washing and recovering the gold in the summertime. The ground was all permanently frozen, so in all the sinking, drifting, and mining the material had to be thawed, generally with steam, in pipes driven into the frozen material and left for a few hours until the gravels had thawed enough to be excavated. The boilers for making the steam were fired with wood, and the hills close to the creeks around Fairbanks were practically denuded of trees. Many of the operations consisted of a group of miners who could raise enough money to lease a property; obtain equipment; buy enough wood, food, and other supplies; and get a cook. They worked on what they called bedrock pay, which was what was divided among the group after the gravel was sluiced in the spring, the gold they recovered was sold, and debts and hired help were paid. One time I asked an old-timer, Andy, where he had met his wife, and he said that a group he had mined with in 1908 had hired her to cook but had had a very poor cleanup and could not pay her, so he had married her. He and Ida had a very good marriage. They lived in a small but very neat cabin at Ester where Andy worked on one of the dredges, and I often enjoyed stopping by to have a cup of coffee with them. They had a 1926 Buick that they kept in perfect condition in a small garage adjoining their cabin. They would not take the car out in the rain or snow, but when the weather was good they would occasionally drive to Fairbanks, about 10 miles, to see a movie or do some shopping. As soon as they arrived home, they would wash the car, and every other month they polished it. When I left Alaska in 1953, I stopped to say good-bye to Ida and Andy and their car still looked like new.

The going wage in 1908 during the early days of drift mining was five dollars per 10-hour day with room and board, which was the same wage I received when I worked in the mine on Chititu Creek 1931 to 1934.

The Fairbanks area in interior Alaska has a very dry climate, warm in the short summers but very cold in the long winters. There were a few hard rock mines that worked all year, but most of the mining was dredging, which could only be done during the warmer weather, so there wasn't much work outside in the very cold weather except prospect drilling, moving dredges from one creek to another, hauling supplies, and the like. I was a panner on a prospect drill for several months one winter, and we would heat water so I could pan the material the drill excavated. We had a small wanigan, a little shack, so we could warm up occasionally. Sometimes it would be 40 degrees below zero or colder. The coldest I was ever out in was 64 degrees below zero, which I experienced when the bus broke down on the way back to college from Fairbanks one night; I had to walk two miles, and believe me it was cold. When you took a car out to a party or dance when it was 40 or 50 below zero, which was not uncommon, you would leave the engine running and at least once during the evening drive to a gas station to fill up. Unless the tires were fully inflated, they would freeze with a flat part on the bottom so you would thump along until the tires warmed up. In recent years, cars have been provided with head bolt heaters, and one can hook up to electric outlets on the meter posts, which is a great improvement in lifestyle in the cold weather.

The Chena River runs through Fairbanks, and there was one small bridge across during the years I was there, so during the winter a ramp would be prepared and a path kept open across the river ice for an extra river crossing. Supplies to isolated mining areas, as well as all the preliminary equipment for developing the Arctic oil fields, were moved by tractor trains or trucks after some leveling across the frozen tundra and rivers during the winter. Two dredges were moved overland during the time I was dredge engineer and later dredge superintendent. One was moved from Goldstream over Gilmore Dome to Fairbanks Creek, and the other was moved from Cleary Creek over the hill to Eldorado Creek. The dredges were cut into three sections, which were mounted on sleds and pulled by large tractors. Where the hills were very steep pulleys were used, so the tractors would be heading downhill while the dredge sections were moving uphill. The steel runners on the sleds had gas burners inside as the runners would quickly freeze to the road if we stopped and had to be thawed loose to get moving again. After I left Fairbanks, they moved one dredge intact from Ester Creek to Sheep Creek over a fairly level area with 15 large tractors pulling it.

The dredges would operate in the fall until the ice in the pond got so thick that the dredge could not move. Then repairs would be done so that the dredge would be ready to operate in the spring. The weather would start warming up by late March or early April, and the winter frost, which would be five or six feet deep, was thawed with steam points. The ice in the pond, which would be three or four feet thick, was cut into three- by six-foot rectangular blocks with a steam cutter, which was a three-quarter-inch pipe with holes in the lower section for steam. The blocks would then have a chain thrown around them and be hoisted on a high line and then released from a cable to a pile outside the dredge pond area. When the dredges shut down in late October or early November, the crews would be let go, except for one watchman for each dredge. Then in March we would start hiring the crews again. Prior to the war it was not difficult to get the men back, but when operations resumed again after the war it was difficult to get enough men for the seasonal work, as more of them could work year-round on construction, mostly for the military bases. We would advertise for men within Alaska and in Seattle and sent recruiters into the northern Eskimo villages to hire men there, too. The Eskimos were generally good workers and could be trained as dredge operators, but they were not very dependable. When fishing or hunting season came along, they would abruptly leave and be gone for two or three weeks. They were a problem on payday as well. Liquor was not allowed in most of the villages, so the Eskimos would go wild when they got some money in the towns. I would go around town with my pickup on Sunday evenings after payday and pick up drunken Eskimo workers around the saloons or in cheap rooming houses and pick up some of the other workers in the prostitutes' cabins and take them out to the camps in order to have a crew to start work Monday morning.

We had some interesting applicants respond to our advertising. I particularly remember reading an application just after I had seen a poster in the post office on hiring-the-handicapped week. The young man said that he had lost an arm above the elbow when in his teens but had been doing all kinds of work and desperately needed a job, so I hired him to work on the cleanup crew. He was a very willing and capable worker and had no trouble keeping up with the other men. His one arm was stronger than two arms on most men. Another applicant was studying mining at the University of Washington. He was from Ethiopia: the government there wanted to train some men for mining in their country, so over 2,000 high school students were tested and the top five were sent to the United States to study mining. The one who wrote to me, Bekele Mammo, placed second in the group and was in his second year of mining engineering and wanted to get some practical experience. At the bottom of a very well-written letter he said that as an Ethiopian he was black and hoped that this would not be a problem.

I hired him, and he was indeed black with perfect features and was a very handsome man. He was also a very good worker and got along well with everyone. I heard about him several years later from people at the Ethiopian exhibit at Expo 70 in Osaka, Japan. They said that Bekele was the deputy director of mining in Ethiopia and was very highly respected.

Audrey and Ted Loftus were neighbors of mine at one time in Fairbanks, and Ted's brother, Art, came up from Oregon for a visit one summer. He would often stop by and chat while I was working in the garden in the evening. After he left me one evening, Audrey came over. She asked if Art had said anything about how long he was going to stay, as she was getting concerned that he was going to stay all summer. I told her that just that evening I had asked Art about his plans, and he had said that Audrey was setting a mighty fine table and that he was enjoying excellent food and good company, so he would be staying for some time yet. Audrey said, "Well I'll fix that in a hurry." Three days later Art was on his way.

Friends from Dawson told me about men they knew during the early years in Dawson who were mining a small area not far from Dawson. They would go into town on Saturday evening to visit with friends at the saloons. One of the men was drinking more and more each weekend, and his partner was having increasing trouble getting him to go home, so often he would go home alone and his intoxicated partner would stagger in much later. One night the one who went home first was awakened by the door being slammed open and thought that his partner must be really drunk, but when he looked around it turned out to be a bear that had broken in. He climbed up on the rafters, went over above the stove, reached down, and got some kerosene and a match. He poured some kerosene on the bear and then dropped a lit match and set the bear on fire. The bear charged out the door and down the trail. The man then jumped down, straightened up the cabin, and got back into bed. Very soon his partner staggered in and said that a bear on fire was charging down the trail and had almost knocked him down. His partner said, "Joe, I've been telling you that you've been drinking too much, and now that you're seeing crazy things like a bear on fire it's time for you to quit drinking." My friends said that the man did cut down on his drinking and always went home with his partner after that.

The Pioneers of Alaska were organized during the early years of the development of mining in interior Alaska. Most of the men who went up during the gold rush or soon after had left family and friends behind, and the Pioneer lodges were organized as a fraternal organization for friendship as well as for helping each other out, burying those who died, and settling their affairs to send what was possible to family far away. At the time I joined shortly after returning from World War II one had to have been 30 years in the territory to join, so I joined Igloo #4 of the Pioneers of Alaska when I was 32. A year or so later I was approached by two past presidents of the lodge who asked if I would stand for election as an officer of the organization. I said that I was undoubtedly one of the youngest members but asked what attributes they were looking for. They said that most members were quite old, many of them hard-of-hearing, and with very little patience, so what they were looking for was someone who could speak loud and little. They said that they knew that I could speak loud enough, and if they could just teach me to speak little enough, I would be all right. I promised to always speak briefly in a loud, clear voice and was elected as chaplain as the first stage of a five-year program, ending up as president in 1952. Elsewhere in these memoirs I mention Eva McGowan and the burial services and the big celebration of the 50th anniversary of the discovery of gold in the Fairbanks area. The officers attended most of the funerals, and one year we averaged almost one a week. Even many members who had families would request burial by the Pioneers. The largest funeral I conducted when I was president was for the president of the local bank, who had requested a Pioneer service. There were over six hundred people filling the local theater. Most often, though, there were only a few people attending, and several times I had to go out to the street to snare enough men to be pallbearers. One time I recall asking Father Mac, who was out for a walk, and even then there were only four of us. Early on someone had written some nice prose for the services. I don't remember all of it now, but one part I always liked went, "His snowshoes are unstrapped, his parka doffed, the storm that wracks the wintry sky no more disturbs his sweet repose, he rests from life's weary journey." During the very long, cold winters we would organize to ensure that someone checked daily on old-timers living alone. Just to see that a light was on and smoke coming out of the chimney was enough. Many lives were saved over the years with this program. At that time the territory had only one Pioneer home in Sitka for taking care of old-timers who could not take care of themselves. It was very difficult to get some of the men to leave Fairbanks and go south, but once they were settled in the Pioneer home they would send word that they were glad that we had practically forced them into going. Now there are state-run Pioneer homes in many towns, and there are people waiting to get into them. I often acted as executor of the estates of some of the old-timers, and it was difficult to locate relatives as most of them had not been keeping in touch with family for many years. It was also difficult to find out what they had, as they generally had their money or bonds hidden in their cabins and very little in banks. One time I was about to donate an old suitcase belonging to a pioneer who had recently died, when another old-timer asked to borrow it; I was cleaning it out and noticed a small slit in the back of the case and found several thousand dollars in bonds. It was gratifying when I did locate more money than one would expect and then locate a relative to give it to. I worked closely with a lawyer whom I had known as a kid in Cordova. He studied law by correspondence and was doing well in Fairbanks. He suggested that I study under his guidance, which I started, but then went to South America and had to stop. I have often wished that I had studied law, however, as much of my work involved legal problems on contracts, titles, and many other negotiations, as well as labor problems.

Gold was discovered in the Fairbanks area in 1902 by an Italian, Felice Pedroni, known as Felix Pedro. In 1952 the Pioneers put on a town-wide celebration to commemorate the 50th anniversary of the finding, and it was indeed a major event. As president of the local lodge of the Pioneers that year I was chairman of the project, and it was a big job. The Italian government sent a bust of Pedro that is now in the university museum. We had a bronze plaque made in Seattle; it was mounted on a large boulder that I found at a dam construction site and we had the F. E. Co. construction crew set it up at the Discovery claim on Pedro creek. A celebration called Discovery Day, with parades and so forth, is still held each year, and the monument on Pedro Creek is rededicated each year. I gave the dedication speech in 1952 and several years ago was invited back to give a rededication speech.

The Italian consul general in San Francisco, Baron Felippe Muci Falconi, came to Fairbanks with his wife Marian for the Discovery Day celebration along with other Italian officials and friends. They were entertained royally, with many parties and a trip to the oil field discoveries in the northernmost

Four old prospectors who knew Felix Pedro at the dedication of his monument, 1952.

part of Alaska. They were wonderful guests and were liked by everyone. A few months after their visit to Fairbanks I called Baron Falconi and Marian when I was in San Francisco, and they invited me to dinner at their home. They had met while they were both studying in Paris when they were young. She was from England and he was from Italy, and French was their common tongue. She taught him English and he taught her Italian. Several years after they were married and he was in the Italian diplomatic service they were assigned to Honduras for several years, where they both learned Spanish. They had four sons and were determined that the boys would learn the four languages that they spoke so they rotated languages at mealtimes. The evening I was there Marian explained their program about languages and said that she would not waiver from the program and hoped I would not mind if they were not speaking English at dinner that evening. Of course I said I wouldn't mind, and it turned out to be Italian that evening. After we were seated, Marian spoke to one of the boys in Italian and he answered in English. She stood up, reached across the table, smacked the boy hard, and said, "When I speak to you in Italian or any other language you know that you are supposed to answer me in that language." He started to explain in English that he had spoken English in deference to me, so she smacked him again and said that she had explained their program to me beforehand. I was somewhat embarrassed, and the boy was angry, but Marian was a determined and disciplined woman and there wasn't anything that either he or I could do. Several years later when I was in San Francisco on business I looked up a friend of the Falconis who had visited with them in Fairbanks. He told me that although the Falconis had been reassigned elsewhere two of their sons were still in school in San Francisco and that one of them was coming to visit him at dinner, so I got to see him. It was the one his mother had smacked that evening. He said that he had been very upset and mad at his mother at the time, but he said, "I really love her now for her discipline as I am graduating next month, fluent in four languages, and I have just accepted a good offer in the Italian diplomatic service."

In the fall of 1933, when I was enrolling at the Alaska Agricultural College and School of Mines (which in 1935 became the University of Alaska), Harold Gillam gave me a ride in his amphibian plane from Cordova to Fairbanks. We landed on the Chena River in the center of town, and Harold invited me to have a good dinner with him before I walked the four miles out to the college with my duffel bag. We went to the Model Café, which faced out to the river, and I had the good fortune to sit at the counter next to Eva McGowan, who became my lifelong friend. Eva had come over from Ireland as a mail-order bride in 1914 to marry a man named McGowan, who was a partner in the restaurant that later became the Model Café. McGowan became sick five years later and was an invalid for a decade before he died. By the time I met Eva she had become a permanent fixture in Fairbanks as the town's official greeter, chamber of commerce representative, and official housing hostess. She was also active in most social activities in Fairbanks. She always dressed in the old-fashioned style with long skirts, high-button shoes, a big hat and scarf, and a fancy handkerchief tucked up her sleeve. Her office was a desk in the Nordale Hotel lobby, and she was known as the "living leprechaun" who put the heart in the Golden Heart City, as Fairbanks was known. Eva was a character in the novel *Ice Palace* by Edna Ferber, and she once appeared on the television show *This Is Your Life*. Every time I went into town from the college I would always stop to visit with Eva and catch up on all the local gossip, which she was always eager to dispense. I worked in Fairbanks for several months after I graduated and would see Eva almost every day, and we often had dinner together. When I left to go to the Army Air Corps on December 8, 1941, it was a tearful departure. Eva did a great service during World War II, when

Fairbanks was so crowded with military personnel as well as civilians involved in construction and support of the army. Many homeowners opened their doors, as did the university, which had very few students during the war years; all of this was coordinated by Eva. According to a story about her life published in *Reader's Digest* she helped about 50,000 new arrivals find housing between 1940 and 1951. After I returned from military service I became active in the Pioneers of Alaska, as I have mentioned previously. One of the important functions of the Pioneers was the service and burial of deceased members. Eva was the most faithful attendee at the services, which averaged nearly one a week the year I was presiding. The president of the lodge conducted the services and always gave a eulogy of the person. It was often difficult to get much information on many of the old-timers as they had been out in the mining areas, and even when they moved into town they were quite reclusive, but I would always find out enough or make up a story to have something to say about them. One evening as I stepped down from the podium Eva approached me and said, "Pat, you should have been a priest; you do such a great job at these services." I thanked her and said that her consistent support made the job much easier and that I sincerely appreciated her constant interest. She then said, "But I sometimes wonder how you can find so many nice things to say about some of these old bastards like the one we are burying tonight." Obviously the candidate that evening was not one of her favorites.

As I have already mentioned, gold was discovered in the Fairbanks area in 1902, and at a Pioneer Lodge meeting early in 1952 it was proposed that the Pioneers put on a celebration of the 50th anniversary of the discovery; as president of the lodge I was to be the chairman of the event. I was overwhelmed at the thought of the responsibility and my lack of experience in such things, but I went to see Eva and she assured me of her support. She helped me outline the committees and the members of each committee I should set up, as well as the newspaper reporters I should contact for their support, and I was off and running. I visited Eva almost daily during the planning and preparation, and the event was a great success. I could not have done it without her support and her intimate knowledge of when, where, and from whom to get the necessary cooperation.

I left Fairbanks in 1953 but returned often for brief visits and always enjoyed seeing Eva again each trip. Soon after Sandra and I married we went to Fairbanks, and the first person we went to see was Eva, who greeted Sandra like one of her own. She gave Sandra a very pretty handkerchief and at every visit thereafter would always give a small bottle of cologne or perfume or some other gift, as well as many kind and loving words. During one visit a few years later we attended commencement exercises at the university, and Eva was overwhelmed when it was announced that a new music room was being named for her in recognition of all that she had done for the university and the community over the years. There was a standing ovation for Eva and not a dry eye in the place, she was so universally respected and loved.

Eva lived in the Nordale Hotel for over 50 years in a small room with a sink in the corner; the bath and toilet facilities were down the hall. She took me to her room once to pick up something, and I could not believe how small it was and how crowded with her things. She would move things from the bed to the floor to go to bed and then move them back to have room to get dressed. A few years ago when Eva must have been 90 or more there was a fire in the hotel. Eva awakened and started going down the halls waking people from room to room but then apparently became disoriented and went the wrong way, was overcome by smoke, and died in the fire. She was praised and credited with having saved several lives. What a tragic end to such a wonderful person, but in a way it exemplified what her life had mostly been about—helping others.

Shortly after the Discovery Day celebration in 1952 Earl Beistline, a former classmate who was then dean of the School of Mines, suggested that I write a thesis to complete the requirements for an Engineer of Mines degree. I got busy and worked very diligently over the next six months. I was dredge superintendent at the time, so in addition to working on my thesis that winter the staff and I were busy planning the rest of the life of the mining operations. The study showed that the remaining reserves would be finished in 10 years, that is, in 1963. I was 38 at the time, and it occurred to me that 38 might be a better age to be looking for a job than 48, but I suppressed the thought as I liked my job and the activities in which I was involved, and most of my family was still in Alaska. However, a week or so later I stopped to see the dean about my thesis, and he said it had been approved and that I would get a degree at the commencement in May. We were walking down the hall after our meeting when he stopped to talk to someone. A bulletin board was close by, so I looked at it and saw a letter from South American Gold and Platinum Co. in New York, saying that they had openings for junior mining engineers for dredge operations in Colombia, South America. When I inquired, the dean said that all of the new graduates were already lined up for jobs and that he was going to advise the company that afternoon. On the spur of the moment I asked him to mention that an experienced dredge engineer might be interested. I soon received a telegram asking for more information, then a request to go to New York for an interview and was offered a position as chief engineer at their principal mine in Colombia. While considering the offer I happened to have breakfast with Dean Patty, who said that it seemed to him like an opportunity for which I had been preparing for many years; he urged me to accept, which I did. I have never regretted the decision, although it was quite a change moving from Alaska to the hot, humid jungle and there were many problems to be resolved, but before I realized what was happening I was in a small, isolated mining camp in the jungles of Colombia called Andagoya.

When I started working in a mine in 1931 at the age of 15 the price of gold was $20.67 per ounce. In 1935, when I was working for the F.E. Co. in Fairbanks, the price was increased to $35 per ounce; this proved to be a lifesaver for the F.E. Co., as it had made a tremendous investment with six dredges and all the supporting equipment of camps, shops, a power plant, and a 90-mile ditch to bring in water for stripping and thawing ground. Costs had increased to the extent that operations were no longer profitable at $20.67 per ounce, but at $35 per ounce the operations were very profitable. However, by the time I left the company in 1953 to go to South America costs had again increased to the extent that the operations in Fairbanks were only marginally profitable. In Colombia the mines (except for the one I went to work for) were doing well with gold at $35, as the local currency, the peso, was devaluing regularly and faster than costs were increasing.

The Colombian company was called Compania Minera Choco Pacifico and was located in the Choco region of Colombia. The headquarters camp was at Andagoya, at the junction of the San Juan and Condoto rivers, which emptied into the Pacific Ocean north of Buenaventura. It was a very hot, humid climate with an average of 270 inches of rain each year; temperatures varied between 80 and 92 degrees Fahrenheit. I had been told in New

York that the company was losing money, but they believed that with technical help the company could be profitable. As I visited the operations and the camp I was appalled by the living and working conditions of the workers, and I soon realized that the basic social problems of illiteracy, poverty-level wages, inadequate housing, extremely poor medical attention, excess personnel, and so forth had to be resolved before the operations could be restored to a profitable level The conditions had been badly misrepresented when I accepted the job in New York, and things were so deplorable that I would have left the first week if there had been a flight out. The weather was so bad that it was ten days before a plane came in. By then, however, I could see a real challenge, so I stayed.

Most of the men were working barefoot, and I was surprised to see that some of the welders had leather jackets and aprons but were welding without shoes. Many of the women were bare-breasted. There was a hospital but it was in poor condition. It had three doctors, as mandated by law, but lacked equipment such as an X-ray machine, a laboratory, and proper equipment. The doctors were treating people with a variety of very expensive drugs, hoping that one of them would work. They had no conception of industrial medicine. Malaria and other tropical diseases were prevalent. Although most staff houses were well screened and had small electric burners and iceboxes, the workers' houses were in deplorable condition. They lacked adequate screening and were equipped only with woodstoves; people had to use outhouses and community showers. Movies were shown on a canvas hung in the street, but the only seats were rocks and it rained almost every evening. A school with three years of education was mandated by law but consisted only of one dilapidated building with a few chairs and benches. Three boards were painted black but they had no chalk; nor did they have paper or pencils. The two "teachers" could barely read or write. A priest came in occasionally from a nearby town and held mass in the so-called schoolhouse. Very few people were married and many workers had two women and often more, so the population increased rapidly. Most houses had at least five children and many had eight or ten. Hardly any of the workers could read or write, so a thumbprint was used on all documents, including pay slips. No one in the area had any knowledge about identifying thumbprints but that seemed of no concern. The company had a store in filthy condition with some staples, but the main items the workers used, such as rice, plantains, bananas, yucca, vegetables, and so forth, came up the river in canoes by farmers who lived downriver. The low wages they received were hardly enough to feed a family of four or five, and most of the family groups had many more members than that.

The managers, engineers, dredgemasters, and electrical and shop supervisors were foreigners. The doctors, a lawyer, the accountants, the warehouse supervisors, and some secretaries were Colombians brought in from the cities. The local workers were Negroes who were descendants of slaves who had run away from their owners in the higher-lying areas many years before and gone to the Choco, which was all mosquito-infested raw jungle. (It had been cheaper for the owners to buy new slaves than to try to recover the ones who had escaped to the Choco.) The Spanish had developed the Choco somewhat for its precious metals, and then the British and American companies had continued the development in the 1800s and early 1900s. Negroes still make up the majority of the population in the coastal areas of Colombia. Andagoya, with its very hot, humid climate and very heavy rainfall, was once described as a place where the flowers have no smell, the birds have no song, and the women have no virtue. Women would go around knocking on the single men's doors at night asking, "Quiere mujer?" (Do you want a woman?) The men said that they paid very little for the services (one man told me that one time he had paid 50 cents and a box of matches). Some men had children by these women, and most abandoned them when they left the area. A few men, however, sent their children to school or at least assisted them. There were many more employees than necessary because over the years the single managers or superintendents had had affairs with underage girls and then hired the girls' relatives to avoid legal problems. The company was the only employer in the area, and there were always many men looking for work. There were 600 employees, but I estimated that 300 employees plus contractors for clearing land would be sufficient if the health and living conditions were better and there were not so many off work from illness.

I had been hired as chief engineer with the understanding that in a year I would be familiar enough with the language (Spanish) and the operations that I would take over as manager from the current manager, who had been there for many years and was ill and wanted to leave. In the meantime I was to do everything possible to make the company profitable. There was much that could be done to improve the efficiency and economy of the company, but it all depended on improving the health and living conditions of the workers and paying them enough so they could eat properly, and then reducing the number of employees. All of these things required an infusion of capital, which the parent company in New York was not willing to provide. I wrote frequent reports to New York advising the actions that were necessary, but every time they wrote back saying that when operations were profitable they would go ahead with my recommendations. This went on for a year and then I gave 30-day notice, as I could not see any hope of accomplishing the things that I thought essential for progress.

At the same time that I arrived in Andagoya as chief engineer a young man named Ray Dunn arrived from New Zealand as mechanical engineer. There was already an engineer from Australia, John Rhodes. The three of us became good drinking buddies, and it was customary there for the staff people to change into their whites after work and enjoy cocktails on the porches of their houses, which were screened all around with long, overhanging tin roofs. It rained very heavily every evening starting about 4:00 p.m., so it was very pleasant having a drink on the porch while the rain cooled things down after having such hot, humid weather all day. Ray, John, and I usually had two or three rum and sodas, then went to the dining room for dinner. One night we drank more than usual, and I did not want to appear in the dining room after drinking too much, so Ray and John went alone. John got into an argument with the Italian maitre'd, who stormed out of the dining room claiming that John had insulted him. He returned a few minutes later pointing a gun at John, just as Ray was coming out of the men's room. Ray was a former rugby player and he made a flying tackle at the maitre'd, who swung his gun around and fired at Ray. The bullet grazed Ray over his right ear, taking the skin and hair off, and others subdued the maitre'd before he could shoot anyone else. Ray quit and went back to New Zealand, the maitre'd and John were both fired, and I was left without my drinking buddies. However, I eventually became president of the company.

The man who owned the controlling stock in South American Gold and Platinum Company, Sam Lewisohn, had recently died, and his widow had decided to sell the stock. Lew Harder, a broker in New York, placed the stock with business associates and family. Lew and two associates visited

Andagoya shortly before I was to leave to see what they had bought. The manager asked me to show them around, even though I told him I could not help but be critical of the New York management. For three days I showed them around the operations while saying what I thought should be done. They went back to New York, had a proxy fight, and took control of the company. They then asked me to move to New York as vice president in charge of the operations, with a free hand to do what needed to be done. It seemed to me that I was going from one jungle to another and that a new suit with wingtip shoes was a far cry from snowshoes in Alaska. I would never have left Alaska to go to New York, but after a year in the jungle in Colombia I was ready to take on another type of jungle. I had never felt small before as the natives in Alaska as well as in South America are all short, but I felt like a midget at 5 feet 8 inches as everyone on the board in New York was over 6 feet and two of them were 6 feet 7 inches.

Orders were immediately placed for electric or kerosene stoves, refrigerators, and toilets for all the company houses that did not have them. Lumber, screening, paint, and the like were made available, along with help for the workers to make their homes secure against insects. Social workers were brought in, and many sewing machines were purchased to teach the women to make clothes for their families. Boots or shoes were supplied to all workers. Wages were increased so that workers could buy enough food for themselves and their families. Two Colombian doctors who had been trained at the Mayo Clinic and were practicing in Bogota were contracted to advise on the renovation of the hospital and the acquisition of supplies and of laboratory, X-ray, and other equipment. A program was set up to send the doctors one at a time to the United States for a one-year course in industrial medicine. I soon learned that it was a mistake to send a doctor alone for a year as there were cultural problems within the families after the man became accustomed to life in the United States, so we sent the wife along and relations were much better. Shortly after the laboratory equipment and a technician arrived there was a noticeable reduction in drug costs, and as living conditions and medical treatment improved it became possible to start reducing the number of employees. Meantime test work had commenced on the dredges to improve recovery of gold and platinum. Bucket line speeds were increased to improve production. Prospecting drilling programs were started to develop more reserves and better delineate existing reserves. Morale of the employees was greatly improved when I issued an order to the effect that training of locals would commence to replace foreigners wherever possible. The gold-dredging operations in the Yukon, in Alaska, and in California had been training operators for many years, and it was relatively easy to obtain experienced men, but this became more difficult as those operations reduced their activities and dredges closed down. I could see the potential of many of the local workers while I was working there, and I thought it was a poor reflection on the companies that they had not trained local people. I put out an order that we would not hire any more foreigners. It was surprising how quickly local people could be trained and there was never any noticeable loss in production, but there was a very noticeable decrease in costs without all the expense of importing foreigners and sending them and their families on three-month vacations every third year. It was difficult to get local engineers to work in the mining camps as they generally preferred working in lower-paying jobs in the cities than going to the jungle camps, but we managed to get a few. After the living conditions in the camps improved some of them stayed, and it was only a few years before we had one local man as manager, and by the time the companies were taken over by Colombian owners a Colombian engineer was the general manager.

Not long after I instituted the policy of training and promoting local people I was on a visit to Andagoya when we received word that a new governor had been appointed for the Choco department and that he was making an official visit to Andagoya. The company had not had good relations with the Choco department governing group, and I was determined to improve the situation. The manager at Andagoya at the time was an engineer from Montana who had worked for many years at Andagoya and then at a smaller operation further south called Narino. He was moved up to Andagoya to replace the sick manager at the time I went to New York. I discussed with him my interest in improving relations with the new governor and suggested that we have a big luncheon with all our staff and that at the luncheon he sit on one side of the governor and I sit on the other side. The manager said he would not sit down to eat with any black man, governor or not. I replied that the man's color should not make any difference; he was the governor and would be treated accordingly. The manager adamantly refused to sit with him. I said that I knew that the manager had

had a local woman for many years and had children by her who were in school in Montana, so I could not understand his statement that he would not sit down to eat with a black person. He said that was right and that a man had to have some principles. I told him that I could not understand his so-called principles and that they were so far from any that I had that he should pack his bags and leave, as there was no way that he and I could work together. He left for good the next morning. The luncheon was a success and was the start of much-improved relations with the local government. Construction of proper schools commenced and an arrangement was made with the bishop of Istmina to obtain teachers. It did not take long to work up to the third-grade level required by law, and then the company voluntarily announced expansion up to the eighth-grade level. The company also agreed with the bishop to build a church and support a resident priest to encourage workers to get married, which hopefully would help control the population explosion.

Once the children learned to read and write their parents expressed a desire to do the same, so evening classes were set up. They were very well attended, and within two years ball-point pens replaced the pads for thumbprints on the payrolls and other documents. Evening classes were also established for training in carpentry and electrical work. A building was erected for a social club and theater. I was there for the grand opening and was very touched when I saw a large sign saying "Teatro O'Neill," which had been erected by the workers to express their appreciation. When the church was completed later on I heard that they were planning to name it San Patricio, but a saint I ain't so I stopped that nice gesture. Within three years the company was showing a profit, which steadily improved, and the company continued to operate profitably for many years after the changes were made.

Shortly after arriving in Andagoya, where all the workers were black and very few staff employees were white, I met a young blonde woman. When I said hello and introduced myself, she responded in mostly swear words, saying, "Jesus but it is nice to see another damn foreigner in this damn lousy place." She went on to say, with many foul words, that she was from Norway and had been working hard to learn English as their contract was almost up and she and her husband were going to visit relatives in the United States on their way back to Norway. She said that Mrs. Davis was teaching her English, and I found out that Mrs. Davis was an ex-prostitute from San Francisco that a dredgemaster had brought down. I told the young Norwegian's husband that he should get his wife's English cleaned up before he took her to the United States or their relatives would be really shocked.

After living in Andagoya for a year I visited regularly for several years but never took the time to see anything except the mining operations and a few exploration sites. However, after Sandra and I married she traveled frequently with me, and she suggested that we take time to see places of interest on the weekends instead of rushing back to New York, as I had been doing. One Sunday we went several hours downriver in a launch to a native village where the Catholic Church had a convent. The Cholo Indians would bring their children in from the jungle for schooling. Since none of them could read or write they would keep track of the time to return for their children by how many moon cycles had passed, but the nuns said that they often missed a month one way or the other. One of the nuns who had learned their language offered to take us to see one of the tambos where three families lived together. It was fairly close to the river and was about six feet above the river on stilts, with a notched log for getting up and down. The tambo was about 30 feet in diameter with a conical thatched roof that extended well down, so there were no sides. The tambo was divided into four areas, one for each of the three families and one for communal cooking and pressing sugar cane. Each family had very few belongings other than bows and arrows, axes, knives, one or two weed mats on the floor, a hammock for the baby, and a basket for their clothes, which were very few. They lived off the land except for what they could get from bartering in one of the villages where they traveled in dugout canoes. They bartered or sold bananas, rice, and some vegetables in exchange for such things as salt, cloth, machetes, beads, and the like. The women were bare-breasted and wore a small piece of material like a miniskirt. The men wore loincloths and sometimes a vest if they had been fortunate enough to acquire one; all were barefoot. The women all wore woven bands of beads around their necks and presented one to Sandra, telling her that now she could go bare-breasted like they were, but she did not want to join the competition. They showed us where the women delivered their babies. They would hollow out a small pit close to the river so it would be full of water. They would go alone to the river with a machete to cut the cord and deliver their baby squatting over the little pool they had made. They had very few children, and we were told that they nursed the babies for a very long time. They were all quite small but nice-looking and very shy. As civilization encroached they would move farther back into the jungle.

The department of Choco was a very poor department, with very little industry other than the mining operations of Compania Minero Choco Pacifico. The majority of the people lived on agriculture or mining or both. Friday was market day, and many people would pole upriver in their dugout canoes loaded with bananas, rice, plantains, fruits, and vegetables to sell to the people in the villages. The banks of the rivers would be lined with canoes on Fridays. Many of them traveled long distances, so they had to start out very early or the night before.

It was estimated that there were as many as ten thousand people (men and women) making a living mining for gold and platinum in the Choco. Most of them were working with just a batea (a wooden gold pan), but a few of them would impound water with small dams and ground sluice the gravel in the hilly areas. Sometimes they dove into the rivers to get gravel from the bottom, which they would pan to recover the gold or platinum. Often you would see them tie a rock on their backs so that they could get down more easily; it was surprising how long they could stay down. There were buyers in all the villages that bought their gold and platinum. It was estimated by people in the government mines department that the production of gold and platinum by the hand miners was almost the same as the production from the five dredges that were operating. The miners would put the concentrates from their panning into a small gourd and take it home. There they would clean the concentrates, separating the gold from the platinum and the black sands, which they would throw out. Many years ago they thought that platinum was unripened gold or a small amount of silver that had no value, so they would throw it out with the concentrates. This had been going on for hundreds of years, and when platinum became valuable during World War I the village of Quibdo, and to some extent other villages, were dug up and the gravels washed to recover the platinum that had been thrown out for so many years. The towns were then rebuilt. The dredges dug up many pre-Colombian artifacts such as gold fishhooks, nose rings, and the like that the early Indians had lost. The fishhooks had no barbs and were only flattened on the end to hold the line, so it was easy to understand why so many of them had been lost. There were also many old Spanish coins that had been lost after the Spanish conquest. Some of these early relics were donated at times to museums in Colombia, Mexico, and the United States. I had some at home one time that I was taking to Mexico, and Sandra suggested one evening that I try on one of the nose rings to show our children. I got it on all right and then could not get it off. The more we pulled, the more the cartilage between the nostrils swelled, and I was becoming frantic as I had a board meeting early the next morning and certainly could not attend with a nose ring. Somehow very late we finally got it off and what a relief that was.

The hand miners would try to work close to the dredges, but to avoid accidents they were generally kept away from the cables that controlled the movement of the dredge. One day, however, someone on Dredge 3 alerted the local people that there were very good values close to the surface, and several hundred men and women immediately started digging and panning in front of the dredge. The operators had to shut down for fear of killing people and sent word to the office. There were a few soldiers or police in each village, but they would not act unless directed to by the mayor, or *alcalde*, as they are known in Spanish. We dispatched the company lawyers to find the alcalde and ask him to order the police to remove the invaders. After several hours of searching they found the alcalde right in the midst of the invading group, and it took awhile to get him to stop his illegal mining and go to his office to issue an order, and most of the day's operation of the dredge was lost before the miners were cleared out.

In its early years of operation the company, instead of putting in schools for all the children, would select occasional youngsters that they thought were brighter than the others and send them to the city for an education. Most of them became lawyers, and some of them returned to the Choco. There wasn't any other industry in the region, so these returned lawyers would often start suits against the very company that had educated them, and they were a constant problem. When I was put in charge, I immediately stopped the practice of sending students out and started developing schools for all the children, which was a very successful program. However, we still had a problem for many years with lawsuits of all kinds from lawyers who had been educated by the company and did not have anyone else to sue.

On one of my frequent trips to Colombia I was rushing to catch a plane from Bogota to Medellin, and as I ran into the terminal about 2:45 p.m. I asked the agent at the counter, "If I run quickly, can I catch the 3:00 p.m. plane?" She said, "No, but if you walk you can catch the noon flight."

The smaller operation at Narino, in the southern part of Colombia, could only be reached via Tumaco on the coast or Ipiales in the interior. Either way required a long eight-hour trip by car to Barbacoas, a small village on the Patia river, followed by an hour or two (depending on the conditions) by launch to Mongon, the mining camp. I have been there when the river rose as much as 15 feet overnight and have had some wild trips on that river. On two occasions I woke up in the morning to find the water up to my bed in the guest house, which was on a hillside above the river. Sandra went with me once each way, but she did not care much for either trip. The road down the mountain from Ipiales was treacherous and narrow, with steep drop-offs, and there was only one place to get something to eat. It was a truck stop, not much of a place, but it usually had something stewing and the food was as safe as one could find in those areas. Tumaco is a small, very dirty port with most of the houses on stilts, so garbage is just thrown out in the hope that the tides will carry it away. However, this doesn't always work, and the place smells terrible, although some people with a good sense of humor call it the Pearl of the Pacific. The eight-hour trip from Tumaco to Barbacoas was very hot and humid and not as exciting as coming down the mountain from Ipiales, but there wasn't any place to eat, so we would take along a tin box of saltines and a can of sardines with a bottle of Aguardiente to provide sustenance and relaxation for the long trip.

There was a big sign on a funeral parlor in Bogota, "Servicio Permanente," which I know meant that they were open round the clock, but I don't know of any more permanent service than a funeral place.

At one time local politicians, many of whom were lawyers that the company had educated, petitioned the government to nationalize the Choco Pacifico Company. The minister of mines and several aides visited the operations to hold public hearings on the proposal. After the local politicians had their say it seemed that the minister would side with them, but then one employee stood up and gave a very stirring speech about how well dressed and housed he and his family were. It was just a few years ago, he said, that all of them had been barely existing, living in rags in a grass shack along the river, and if the politicians had their way they would destroy the company with graft, corruption, and poor management and the employees would once again be barely existing in rags in a grass shack watching the river flow by. His speech carried the day and the government took no action on the proposal to nationalize the company, so it continued operating. However, about 20 years later, after the nationalization of major mines in Chili and other countries, a proposal to nationalize all the mines in Colombia without compensation was presented to Congress. It seemed very probable that such a proposal would succeed in Congress, and I was worrying about what we could do when I happened, on a flight from Medellin to Bogota, to sit beside an astute, well-connected lawyer who had been a minister and ambassador. I explained our problem to him and asked if he would be willing to help us. He thought that he could, so I put him on the payroll in Colombia and then we went to New York to negotiate a deal with Lew Harder and the board. We made a deal with the gentleman and agreed to pay him a percentage of any funds received in dollars in New York from a sale of the properties. He first arranged with the president and leaders of Congress to put the proposed law on hold to give us time to negotiate. Then he sought the support of ANDI, the Association of National Industries, convincing them that if the government nationalized the mining industry it would be but a short step to doing the same to other industries in the country. He then got a group of businessmen and bankers together and convinced them to buy the mining companies, which they did, except for the Frontino company. A reasonable sale price was arrived at, and the funds were remitted to New York. The auditors at year's end were sure that we had paid off or bribed someone but were finally convinced that the deal with the negotiator was a legitimate business deal. I never knew whether or not he had paid someone, nor did I care, since he got the job done that we had contracted with him to do. It was interesting working with such a well-connected lawyer: it would normally take a week for me to get an appointment with the minister of mines and longer for one with the president. In the morning the lawyer would call the president or the minister of mines or any of the bankers we were negotiating with and invite them for lunch at his elegant home, and they would be there, so the negotiations moved right along.

Bribing or paying people off was very common in Latin America, but we never did it during the many years that I was in charge of the operations. We did hire a lawyer occasionally when we had to get something done quickly, such as getting papers legalized or urgently needed equipment out of customs, and we never knew whether or not the lawyer paid anyone off and we were not concerned, since it was just a business deal to get a job done. In Bolivia, for many years before we went there, it had been the policy of the major mining companies to give money regularly to government officials; I had many problems until they became convinced that we were not going to pay them. Each time the government changed, which was fairly often, the new group of officials would not believe that we had not been paying off the previous group, so we would go through the same problem again. One time at a party at the mining camp a new minister who had been drinking heavily pointed his gun at my stomach and said that if I did not put him on the payroll immediately he would pull the trigger the next time. I finally convinced him the next day that we had not paid anyone and we were not going to, so he stopped pestering me. He was never cooperative, though.

Frontino is an underground gold mining company with about 650 employees and an equal number of retirees. The laws were very liberal about retiring on pension at 75 percent pay after 15 years working underground. In addition, a series of laws had been passed in recent years granting the widow of a pensioner his full pension for the remainder of her life, followed by a law requiring the companies to fund the future pension obligations. Most of the older pensioners had never been legally married but would marry a young girl before dying, and the company was faced with many young female lifetime pensioners. Frontino was moderately profitable, but there was no way that it could possibly put up the millions of pesos required for pension funding. The pension fund was the only liability the company had, and no one would have been interested in buying a company with such a liability, so the company was turned over to the pensioners. The company was still incorporated in England and I was the chairman, so I merely wrote a letter explaining the situation and gave the company to the pensioners, who worked out a deal with the union for a joint venture. At the time of this writing almost 30 years later the company is still operating.

The people in the Choco operation did not fare as well, unfortunately. The Choco operation was marginally profitable and could have continued operating for many years with the conservative management of International Mining. However, the group that took over the companies had a more extravagant lifestyle than we had ever dreamed of. We had had an older Oldsmobile in Bogota with a chauffeur who was not badly dressed (usually because he wore my hand-me-downs). They, on the other hand, soon had three new Mercedes cars with liveried chauffeurs. Then they acquired a much newer and bigger airplane than we had had, as well as larger and better-furnished offices in both Bogota and Medellin. They also increased the staff. Fortunately for them the price of gold went up dramatically, so they were able to absorb higher costs for awhile. In the Choco they had problems with the local people and politicians, though, and soon the new group took out all the company's cash reserves and then turned the company over to the union. The union leaders were soon stealing the production, so there wasn't any money for wages and supplies, and the company soon closed down. It was an unnecessary but very real tragedy, as pensioners who had worked faithfully for many years lost their pensions. The employees, too, lost not only their jobs but their well-earned pensions. I received heartbreaking letters from some of the old pensioners and workers, and occasionally sent some of my own money to help the saddest cases. I tried to do what I could with the people to whom we had sold the company and the government to help the destitute pensioners. The government finally paid partial pensions for a brief time, but it was a very sad ending for a company that had treated their people well for many years and had done so much for the community.

On one of my earlier trips to the Choco operations, I flew to Bogota by way of Buenaventura and Cali. My first evening in Bogota I was invited to a cocktail party at the U.S. ambassador's residence. At one time during the evening, a lady behind me was telling the ambassador what a terrible trip she had had that day from Buenaventura to Cali. She said that she had boarded an old DC-3 that was in deplorable condition, and after waiting awhile a disreputable-looking man had emerged from the cockpit saying that his copilot had not shown up and asking if anyone on board knew how to fly. She then described how an equally disreputable-looking man across the aisle from her had stood up and said that he knew how to fly and had gone to the cockpit. She said that it was the worst trip she had ever made, very turbulent, with very bad weather, much longer than planned, and just a terrible trip. She was especially worried because of the appearance and questionable ability of the two men up front. Fortunately, she finally arrived in Cali and had a nice trip on Avianca to Bogota. I then turned around and spoke to the ambassador, who introduced me to the lady, and I informed her that I was the "disreputable-looking" character sitting across the aisle from her who had flown as copilot. She was somewhat embarrassed but turned out to be the wife of the head of Pato Consolidated Gold Dredging Company, whom I knew, and we became friends. International Mining later bought control of Pato. It really had been a bad flight from Buenaventura to Cali, with very turbulent weather, and we had to fly a long way out of our route to avoid some of the worst of the weather and so had a longer-than-usual flight.

I lived only one year at Andagoya in the Choco as chief engineer before being transferred to the New York office as vice president in charge of operations. At that time the company had only two operations, the one at Andagoya in Choco with five dredges and one farther south in Colombia at Narino with one and later two dredges. Not long after I went to New York the name of the company was changed from South American Gold and Platinum to International Mining Corporation, mainly because every time there was bad news about anywhere in South America the stock price was adversely affected. I usually visited the operations in Colombia once a month, and before the company acquired a plane I would take a commercial flight with Avianca from New York to Cali and from there would take a smaller Avianca plane that went two or three times a week to Andagoya. One time I arrived too late for the Avianca flight to the mine, so the agent arranged for a charter flight. It was a small Cessna with a fairly young pilot. When I arrived at the plane, the pilot said that he understood that I wanted to go to Andagoya, but he did not know where it was as he had never flown in that area. He had been told that I knew something of the area, so if I was willing we could go. So off we went and I told him to fly northwest until we came to a big river, the San Juan, and follow it north until we found Andagoya at the junction with the Condota River. Fortunately the weather was good and we had no problem. The western ridge of the Andes Mountains is between the Cauca Valley (where Cali is located) and the Pacific Coast, and on two occasions planes flying from Andagoya in bad weather had crashed in the mountains with company employees on board.

Newton Marshall was the president of South American Gold and Platinum Company when I joined in 1953. He was a tall man with a deep depression in his forehead; his nose angled off to one side, his mouth and chin were badly scarred, and his mouth opened mainly on one side. It was obvious that he had been in a terrible accident, and I did not know him very long before he told me about it. It happened in 1934, and he was flying from Andagoya to Medellin and Bogota. The plane was a small amphibian operated by a German company, Scatda, with a German pilot and mechanic. Scatda had the main air service in Colombia up until World War II when, because of proximity to the Panama Canal, the United States asked the Colombian government not to allow German pilots to fly, at which point Avianca took over all flights in the country and used American or Colombian pilots. The amphibian took off from the river below Andagoya, stopped at Quibdo, and then took off for Medellin. The other passengers were a German with a small dog and Don Julio, an acquaintance of Mr. Marshall who was sick and heading for a hospital in Medellin. They flew in and out of clouds, circled around some, and then started to let down in the clouds. Mr. Marshall thought that they had gotten over the western range of the mountains and were letting down in the Cauca Valley, but instead they crashed in the mountains. His first recollection was that he had had a bad blow to the head and that there was something in his mouth that he could not spit out; he then realized that it was his upper lip. It had been almost cut off and was hanging by a small piece of flesh. His lower lip was cut down on both sides at the corners and was hanging down on his chin. He got out his handkerchief and held his lips in their proper place as well as he could. He was bleeding from his mouth as well as from his forehead and nose. Don Julio and the dog were dead, so the pilot and mechanic took them out of the cabin. The German passenger had a broken arm and leg, which the mechanic bound up with pieces of wood. The pilot bandaged up Mr. Marshall's lips as well as he could. He could not find any pins, so he used his pilot wings to fasten the bandages. The pilot and mechanic tried to make the two passengers as comfortable as possible in the damaged cabin, both of them still being in their seats, and then they left in search of help. Mr. Marshall was able to drag himself around a bit after the third day, found some paper cups, and was able to catch some water dripping off the fuselage to give to his companion. The man did not speak any English but did know the Spanish word for water, *agua*, which he asked for frequently. Mr. Marshall gave him some whenever it was raining, which was most of the time. However, the man died during the fifth or sixth night after the crash. Mr. Marshall managed to make a hole in his bandage so he could pour water into his mouth but most of it ran out where his lower lip had been severed. He would hear planes whenever the weather was clear, but none of them came close enough to see the crash. He tried to move around more each day and managed to gather some wood so that he could light a fire whenever a plane came close. He told of leaning against a banana tree for many hours while waiting for the sound of a plane, and he said that he could hear the banana tree growing. After 10 days he found a small mirror and then soaked the bandage off his face and was shocked at what a sight he was. His upper lip, which had been nearly cut away, had grown together off to the side of where it should have been. The lower lip was only partly healed with deep cuts on each side where water had run out whenever he tried to drink. Some teeth were missing and others were loose, but he did not have anything to eat anyway so that was not an immediate problem, although he wished that it might have been. His forehead was deeply depressed, and he had constant headaches. He kept hoping that the pilot or mechanic would show up with help or at least that the wreck would be spotted and help would come, but after two weeks he decided that his only hope was to walk out. He had had only water but found that even without food he was feeling stronger, and he prepared to walk out. He climbed up a hill above the crash site and recognized Mt. Torra, a landmark in the Choco, so he knew that he was on the western side of the mountain range and that his best bet was to try to get down to a river where he might find some habitation. While getting some things together to walk out he found a small tin marked "for emergencies" that had only some pieces of chocolate in it. He could not chew the chocolate but was able to let pieces dissolve in his mouth, which gave him some sustenance on his very difficult walk out. His only shoes were dress Oxfords, and he had to cross many streams and very rough terrain, mostly in the rain. He could only travel in daylight and spent many wet hours during the nights. After three days he met some men from Antioquia who were out looking for old Indian graves for the gold ornaments they might contain. They had found the body of the pilot alongside a stream, but no one ever reported seeing the body of the mechanic. The men said that it would be too difficult to continue out to the Sipi River, where Mr. Marshall was headed, and insisted that they go back on the trail they had made coming in. After five days of going up and down trails and crossing streams Mr. Marshall was about ready to give up as his feet were badly swollen and sore, and he did not dare take off his shoes as he knew that he would never get them back on. They finally came to a small farm where they were able to get some horses to take them to where they could get a car and drive to a town, where they got medical attention and sent out news that he was alive. He was indeed very fortunate to have survived such an ordeal for 25 days, the only survivor of the crash.

A few years after I took over the operations in Colombia, an Avianca DC-3 flying from Andagoya to Cali crashed in the mountains. Several employees, including Vince White, the dredge superintendent, were on board, and all were killed. A shipment of gold and platinum was also on board. The natives were the first to get to the plane, two days after the crash, and by the time officials arrived a day later the gold and platinum had disappeared, as well as the jewelry and money of all the deceased. The gold bars were easily disposed of with refiners in Colombia, so we never did find out anything about them, but the platinum could only be disposed of through dealers or refiners in the United States or Europe. The two cans containing 300 ounces of platinum, each with the company seal on them, showed up two or three months later, and the insurance company successfully claimed them.

The loss of valuable employees prompted the companies to acquire an airplane for transporting staff and valuable cargo. By that time, International Mining had taken over the Frontino and Pato operations and therefore controlled all the precious metal mines in Colombia, and I was in charge of the operations of all of them. The plane was very convenient and a great time-saver for me in my frequent visits to the mining operations. However, the local politicians, who were always criticizing the companies in the press, said that Sandra and I were flying out of each camp with a load of gold. Little did they know that when I took Sandra to the labs where they were pouring the gold bars she could not lift one, and I had trouble doing so. International Mining also had an Aero Commander plane stationed in White Plains, New York, and over a period of a few years I made 19 or 20 trips from New York to Colombia; several to Ecuador and Peru, where the company had interests; and several to Bolivia, where we were exploring and where we later installed a dredging operation.

Vince White and his wife were from Dawson in the Yukon, and they had been a great asset to the operations at Andagoya. He was an excellent dredge operator, and she was an experienced nurse who often helped out in emergencies. After Vince's death Florence moved to Redwood City, California, near some relatives. Her father, Mr. Hartley, was visiting her from Dawson, and on one of my frequent trips to San Francisco, Florence asked me to drive Mabel Franklin down to Redwood City so that she could visit with her father. Mabel's deceased husband, Colonel Ray Franklin, had been in charge of gold dredging operations in the Yukon in the early 1930s; he had then gone to Colombia, where he was in charge of the Pato operations during a major expansion program. I have already mentioned Mabel Franklin as the lady who was telling the ambassador in Bogota about her terrible flight from Buenaventura to Cali on which I had been pressed into service as copilot. Mr. Hartley was in his late eighties and very spry. When we were alone, I asked him what he did to keep in such good physical condition. He said, "It is exercise, my boy, exercise. My daughter thinks that I drink too much and am on the verge of becoming an alcoholic, so she will not have any whiskey or beer in the house. She does leave a little money for me, though, and I have a few dollars of my own. The closest bar is two miles away, so I walk there twice a day, have a few drinks, and walk back. So it is exercise, exercise, my boy, that keeps me in such good condition."

The first acquisition by International Mining was to purchase one-third of the stock of Pato Consolidated Gold Dredging Ltd., a Canadian company based in Vancouver, B.C., and controlled by Placer Development, which also owned one-third of the stock. One-third of the stock was publicly held, and the one-third purchased by International Mining had been held by an English company that wanted to sell because of double taxation in England. Pato's operations in Colombia, in Antioquia on the Nechi River with headquarters at Bagre, were larger and newer than the Choco and Narino operations. A modern camp had been built and new, larger dredges had been installed in the late 1940s. Seven dredges were operating very profitably. The chairman of International Mining and I went on the board of Pato, and International Mining soon began acquiring shares in Placer Development, and we both were elected to the board of Placer. Placer had been stagnating under a chairman who had been a great leader but who was now suffering from advancing senility, and the company was adrift. The directors forced the resignation of the chairman and appointed an aggressive president, and Placer started moving forward with the development of several mines. It was an interesting time, and International Mining acquired a substantial block of Placer stock. The Canadians were concerned about a U.S. company taking control, and after long negotiations International Mining sold their block of Placer stock at a substantial profit to Noranda, a major Canadian mining company, and we went off the board of Placer. We suggested that Placer should either buy International Mining's one-third of Pato or sell their one-third to International Mining. The Placer board was in agreement, and the chairman of International Mining suggested that we proceed like children cutting a cake in half—Placer would name a price and International Mining would either buy or sell. I made an evaluation of Pato and came up with a price of $3.50 per share. If they named a lower price we would buy; if a higher price, we would sell. The Placer group came up with a valuation of $3.25 per share, so we excused ourselves for a conference, went to the washroom to kill a little time, and then went back to the meeting and advised that we would buy Placer's share of Pato, which we did. I was elected chairman of Pato, and Placer allowed us to keep our headquarters in their office in Vancouver for many years. Eventually the domicile was moved to Bermuda after the operations in Colombia were sold to the Colombian group and International Mining was taken over by Pacific Holding. However, I greatly enjoyed visiting Vancouver regularly during those years; it is a nice town in a lovely area. At one of the annual meetings in Vancouver, a shareholder asked what the name of Pato was prior to my changing it to mine—Pat O'Neill. I explained that the name Pato, which means duck in Spanish, was the original name of the company, and that I had never thought of the connection to my name.

I had some exciting times with the Pato operation including a holdup and robbery of the cleanup from one of the dredges as it was en route to the camp on the river. There was also a holdup of a gold shipment at the airport in Medellin from the company plane as it was turning from the end of its landing run to taxi to the terminal. An accountant from England, named Edward, had come out to South America to work with an auditing firm, and after a few years in Chile he moved to Colombia, where he worked for Pato. He had been married in England to a cousin, and they had two boys who stayed in England. They adopted two girls in Colombia while Edward was working in the Bogota office. However, when he returned from home leave before going to Bagre, his wife and the girls stayed in England, and they divorced soon after. When I took over the Pato operations in 1961, Edward was the assistant manager at Bagre. We had a German pilot at Bagre who had an affair with their maid while his wife was away on home leave. When the maid became obviously pregnant and the talk of the camp, Edward called the pilot in and told him that he would have to get the maid out of the camp, and set a deadline for doing so. The Saturday night before the deadline Edward was at a cocktail party at one of the staff homes, sitting in an easy chair. The pilot walked in, went directly to Edward, and asked him to extend the deadline. Edward said he did not believe it convenient to extend the time, and the pilot pulled out his pistol and shot Edward. Fortunately Edward had started to get up just as the shot was fired, and the bullet entered just below his heart. The pilot walked out and went to his own house, laid down on the bed, put the pistol in his mouth, and killed himself. Edward recovered and continued working at Bagre.

Not long after the shooting incident the company and the union were negotiating a new labor contract. The union went on strike without notice and blasted down power poles, cutting off all power. One of the dredges had the digging ladder at full depth and it was sanded in before power could be restored. It took two months after the strike for another dredge to dig its way over and release the one that was sanded in. During that strike, meetings between management and the union became angrier and feelings were very hostile. One day after a meeting broke up with very bad feelings on both sides one of the union leaders returned with a pistol and shot the manager, who had just walked into the accounting office. The man shot from outside through a window; one bullet penetrated the joint of the cap on the manager's skull and the other, his groin. The manager recovered in a hospital in Medellin but would not return to the camp, so I named Edward the manager. On another occasion a guerilla group attacked one of the dredges during the night, stole what they could, and then sank the dredge. Fortunately, the crew got away safely by swimming to shore. Security was a constant problem and stealing by the dredge crews was very difficult to control. They would spend more time thinking about how and where they could steal than the operators thought about how they could thwart the thieves.

I had heard that Edward was romancing a social worker, so I was not surprised when he sent word that he would like me to meet his fiancée on my next trip. We arranged to meet in Bogota, and I was waiting in the hotel lobby when Edward walked in with two ladies. One was a lady that appeared to be about Edward's age, and the other was a younger woman who I thought was probably a maid, so I greeted the older one with enthusiasm. I was immediately advised that she was the mother of the younger one, who was in fact the fiancée. It was an embarrassing moment. The young lady soon married Edward, and although she was always polite and stiffly friendly, she never did like me after that first meeting. Edward and his new bride made no secret of the fact that they were trying to have a baby. It was the talk of the camp when the doctor would go to their house at odd hours to give her an injection or whatever to improve their chances of conception. After three years they decided to adopt and made an arrangement with a doctor in Medellin, who knew a pregnant young woman from a good family who was willing to give up her expected child for adoption. They were waiting nearby, and when the young woman went into the delivery room Edward's wife checked into the maternity ward. When the baby was born, the doctor carried it into the wife's room, laid it alongside her, and told her that she had just had a baby boy. The doctor then filled out the papers accordingly so they did not have to go through adoption proceedings. It was a surprising and certainly an illegal arrangement. It worked, though, and the boy turned out to be a fine young man.

I helped a childless young couple from Pato adopt a baby girl in Bogota. The paperwork was tremendous, and we had to get many documents, many of which I had to complete as their employer and guarantor of their character and responsibility. This happened in 1964 or 1965, and I still receive a Christmas card with a nice note and often a picture of their daughter every year. I wrote some letters for another couple, Jack and Merle Swan, who worked with the companies for many years. They adopted a boy in Northern Ireland when they were on home leave in the mid-1960s. Jack was from Ireland and Merle was from Canada, and they were in their early fifties when they adopted the boy. You never saw a couple that doted on a baby like they did. He would ask for something, and they would practically knock each other over trying to be first to get what he wanted. They later worked in Colombia, first in one of the camps, then later in Bogota. When they retired and went home to Ireland, the boy, William, was about 10 and quite spoiled. They sent him to school in England, and when next I saw him about six or seven years later he was the nicest young man you would ever hope to meet. Later on he took wonderful care of his parents in their failing health and married a nurse who took care of them in the hospital before they died. He now has a very successful business in Northern Ireland and has two young children, and we still keep in touch.

When I was in a hospital in New York in 1964 for an operation, there was a pretty little girl about five years old whom I often visited with. She had one terribly enlarged finger that had to be amputated. I called my secretary and asked her to get a really nice doll, which I gave to little Kathy after her operation. I have received a Christmas card every year since from Kathy, always with a nice note saying that she still has the doll and telling me about her family.

The Colombian group that purchased the properties of International Mining changed the name of Pato to Mineros Antioquia, and the company has continued operating up to the time of this writing. Not very peacefully, however, as guerilla activities and kidnapping have increased to the extent that it is not safe to travel anywhere in the country. I did some consulting for the Mineros group over the years, despite serious concern about my safety. On one trip, the guerillas, who had been threatening sabotage to obtain payoff for peace from the company, dynamited part of the office in Medellin the night before my arrival, destroying maps and records that I needed to study. We were to fly out to the camp the next day, but that night the guerillas destroyed the tower at the airstrip and two planes that were on the ground, closing the airstrip. We chartered a helicopter to take us out. During the two days I was in Medellin 21 people were murdered, so I did not go back for several years. In 1995 the manager of Mineros assured me that after the arrest of several of the guerilla leaders the situation had improved and that it would be quite safe for me to visit, as he wanted to discuss some proposed projects. So I went down and took my 23-year-old son Kevin along. When we arrived at the mining camp, El Bagre, I was surprised to see an army encampment of approximately 200 soldiers; they were manning armored posts on each corner of each of the five dredges 24 hours a day as well as conducting armed patrols throughout the camp. A helicopter was used by the gold recovery crew and management to travel between the camp and the dredges, as it was not safe to travel on the river as we had done for many years. The first evening in camp the colonel in charge of the army troops was at a cocktail party at the manager's house. He said that although there had not been much activity in recent days, about three weeks previously the guerillas had attempted to invade the camp and the army had killed 13 of them. He also said that a known group very active in kidnapping had their headquarters in a small town nearby, and since I was well known in the area from my many years as chairman of the company I must be especially careful. When he learned that Kevin had diabetes, he insisted that we not go anywhere without a military escort, as Kevin would not last long without insulin if kidnapped. We visited all the dredges in the helicopter with armed guards, and we went in the helicopter to the headwaters of a tributary of the main river where the company has a hydroelectric plant. There was a swinging bridge across the canyon by the plant, and Kevin wanted to cross it and do some climbing. The soldiers stopped him, however, saying that the hillside was mined because there had been several attempts by the guerillas to invade and that three had been killed in recent weeks. It was a nervous time, and I did my studies of the proposed projects as quickly as I could. While I was working in the office, Kevin spent some time with the army officers, and they took him out to the rifle range and let him shoot all their different weapons, which he enjoyed.

The Mineros group was doing some exploration in an underground mine south of Medellin and wanted me to see it, so after we returned there we went in an armored Chevrolet Blazer, about a two-hour drive. The vehicle was completely armored and had bulletproof windows, only one of which could be opened just enough to slip out a bill to pay tolls. A radio was installed with direct connection to the police in case of a holdup. One of the directors of Mineros, a well-known business man in Medellin whom I had known for many years, went with us to visit the exploration work. On the way out, he told us that his oldest son had been kidnapped three months previously and that he was still negotiating with the kidnappers. They wanted more money than he could arrange, especially since he had eight other children, and if he paid what was demanded the kidnappers would go after them as well. He said that he still had hopes of getting his son back, but when we arrived at the mine, there was a message saying that his son's body had been dumped in the front yard of his home shortly after he had left that morning. It was a terribly traumatic experience for Kevin and indeed for all of us. From Medellin, Kevin and I went with the manager to Bogota to attend a board meeting of Mineros. The company car that met us and took us around was an old beat-up taxi with a chauffeur in very old clothes and a pistol in his pocket. It was quite a change from the days when I had been in charge before the kidnapping and guerilla activities, when we had a good car with a well-dressed chauffeur who wore all my used suits and sport jackets that I would take down. One time when the manager came out to the airport with the company car to meet me, both the manager and the chauffeur were dressed in blue blazers and gray slacks with blue shirts and striped ties. They looked like bookends, and the manager commented that he wished I wouldn't bring such nice clothes down to the chauffeur. I wanted Kevin to see the famous gold museum in Bogota while we were there, so the armed driver dropped us off and picked us up at the front door at an arranged time. While we were in the museum, there was a fairly strong earthquake, with the building shaking and the chandeliers swinging. Many people ran out, but the streets in Bogota are so unsafe that we decided that we were just as well off in the building. It was an interesting and exciting trip. Kevin and I enjoyed it, but I was relieved when we left for Mexico.

An engineer, Carlos Aspillera, who had worked for me in Colombia and Bolivia and whom I often recommended for consulting jobs, was kidnapped while on a consulting job north of Medellin in Colombia. Before I heard that Carlos was kidnapped his wife called me from California crying and saying that she was sure their marriage was over. She and Carlos had been arguing when Carlos left, and for the first time since they were married he had not been in touch for her birthday. I told her that he was probably out in the jungle and could not communicate, which indeed he was. He had been taken to a small camp in the jungle that was in deplorable condition. He knew that his kidnappers had recently received two million dollars from another kidnap victim's family, so he convinced his captors to spend some money fixing up a decent tent camp with some cots and to get a cook and arrange to have food brought in regularly. He thought that they should at least have a comfortable camp with good food during the long period of negotiations. His captors believed that he was working for a large company and were demanding a million dollars. The engineer finally convinced his captors that he was not working for a company and had been examining the mine where they caught him on his own accord, to get information so that he could promote the mine to a major company. After three months, they agreed to let him go for $18,000. However, before releasing him they asked him to join them as their commandante as he was such a good organizer. He declined, but they gave him a letter to show to any other kidnappers who might catch him that he should not be detained a second time.

Colombia went through a very violent period in the 1950s, with great conflict between the liberals and the conservatives. Many people lost their lives, and occasionally a body would appear floating where one of the dredges was working in the river. The workers would secure the body and notify the local officials, who would recover it and take it for burial. These incidents became so frequent that the authorities told the dredge operators to just move the bodies out into the river and let them go. Violence tapered off after a dictator took over the country and restored law and order. Someone told me that one evening during the bad period he was driving into Medellin and picked up a man who was walking with a bag on his back. He asked him what he was carrying, and the man replied that it was a head that he was taking to the man who had contracted him to kill the person, to show that he had done the job.

Pato had an engineering and designing company in San Francisco that did the purchasing and engineering for the Pato operations as well as for the International Mining operations. I was chairman of that company also and visited it often. Sandra and I had spent part of our honeymoon at the Huntington Hotel in San Francisco, and on our frequent trips there we stayed at the Clift Hotel. We found out early on that a person who had stayed at the Clift for years had had one of the rooms enlarged, so we always asked in advance and got that room. We always enjoyed our visits there, especially the beautiful flowers we always had in the room and the many great places to eat. I usually went to mass at Old St. Mary's on the edge of Chinatown on California Street on my way to the office in the morning. I always enjoyed Old St. Mary's and thought it especially appropriate in Chinatown that the statue of Mary had slanted eyes. Just recently Sandra and I were on a cruise that stopped in San Francisco for a day, and it was great fun to take the cable car from the pier to the center of town and have an excellent lunch in the Palm Court in the old Palace Hotel. What a lovely city and great memories!

The next acquisition by International Mining was Frontino Gold Mines, Ltd., an English company that had been operating in Colombia for 155 years when it was purchased in 1956. The English owners had lost interest because of double taxation and had not been reinvesting, so the industrial plant was badly in need of modernization. The staff camp was very nice, with well-built, comfortable houses on the ridges or hills around the mine, excellent staff club facilities, and a nine-hole golf course. The mill was a disaster, however, and the mine was badly in need of more development and better organization. I was elected chairman, and we started an aggressive program to improve all aspects of the operation. The main mine at Frontino was a rather narrow vein that extended as much as 2,000 feet laterally and was down 30 levels (approximately 3,000 feet) along the dip of the vein, which was on a 60- to 65-degree slope. There was a high-grade streak in the footwall of the vein, and it was a constant problem to control the high grading, that is, stealing of the high-grade material. The workers would pick out high-grade material whenever the foremen were not close by. They would put the high grade in a sock or something similar and tie it in their crotches and around their waists. The security men would search them and catch them with the material, but unfortunately the union was able to convince the authorities that the security men were homosexuals and ordered them to stop searching the men's crotches. The manager said that he would hire female security guards but that created even more objections, so they settled that the men would have to strip before going in to the change room so that the guards could see if they had any high grade when they came out of the mine. Men who were not employees would bribe their way into the mine and spend as much as a month or more in the upper old workings. High graders, or *machuqueros*, as they were called, would go down to the working areas during the night or on Sunday when the mine wasn't in operation, to dig out the high grade and concentrate it in makeshift mills. When they had enough concentrates, they would go up toward the surface and dig their way out, so every once in a while you would see a new opening. It was very dangerous to try to catch the invaders as they were in the dark taking turns as watchmen so that they could see anyone approaching with a headlamp and shoot them. I was busy in Bogota one trip but was told that high-grade ore had been encountered in a new drift in the lowest level of the mine. I flew into the camp on the weekend and told the mine superintendent that I wanted to go down to see the new development on Sunday morning. He said that there was only one access to the place as the shaft had just been sunk and the drift advanced only a few feet. It would not be safe to go down on Sunday as the *machuqueros* would be working there, and without an escape route they would kill us. I thought that it was ridiculous that the chairman of the company could not safely visit the mine when he wanted and told the superintendent to get word to the *machuqueros* that I was going down and would have armed security men with me. We got word back that it would be safe for me to go down between 10:00 a.m. and noon, so that is what we did.

The high graders and thieves in the area, and even many of the employees, were dangerous and ruthless. The mine superintendent fired one of the workers who had been caught carrying out high grade, carrying supplies into the mine for the *machaqueros*, and high-grading away from his designated working place. Three days later the man walked into the superintendent's office when he was alone and slit his throat. The killer was in jail very briefly, claiming that he had acted in self-defense, even though he had had a knife and the superintendent had not been armed. Thievery was rampant in the homes and in the surface mining operations. Families would leave security guards at their homes when they went out for the evening, but the thieves would threaten, wound, or kill the guards and steal what they wanted. It was not unusual for the thieves to steal telephone and power cable lines, which they could sell for scrap. The local judge would do nothing to the thieves except give them a light slap on the wrist. In an effort to get some control of the situation, the manager and I went to Medellin to ask the governor to assign some troops to the area. He refused, saying that there were many poor, unemployed people who had no alternative to stealing. We told him that we agreed in principle with him but that we could not protect our property and our employees, so we would provide transport for the thieves from our mine area to the area in Medellin where he lived where they could undoubtedly find richer homes to steal from. So he agreed to send out some troops and a new judge, and the situation at the mine improved.

Many years earlier, as part of a labor agreement, the company had agreed to provide a more-than-generous quantity of meat each day to the workers and their families at very little cost. So the company developed a ranch that had 6,000 head of cattle at the time International Mining bought control of Frontino in 1956. The manager at the time was a crusty old British miner who loved the ranch and the horses. When I arrived in camp as the recently appointed chairman, the manager took me around, and we got on quite well in the mine, the mill, and the camp facilities such as the hospital, school, commissary, housing, and so forth. He then said that he wanted to show me the ranch, which he described as the showcase of the operations. Early the next morning the manager and I met with the ranch superintendent and two of his foremen and horses at the end of the road. I had only been on a horse once in my life, when I was 17 and the bridge had washed out on our way to the Chititu mine in Alaska in the spring. Each of us had been boosted on a horse, and we had forded the river. Now, after many years, I was on my way to go horseback riding, and I recalled hearing that there was a right and wrong side on which to mount a horse. I did not know which was which, but I did not worry about it as I figured that I could observe the others and follow suit. However, when we got to the horses, the manager said that it was proper etiquette for the chairman to mount first. I insisted that the others go ahead as I had to go to the bushes for a few minutes, but they all insisted on waiting for me, so I finally went ahead and unhappily mounted on the wrong side. The old manager was terribly exasperated, threw his sombrero on the ground, and said that after 30 years managing mines he now had to work for an ignorant so-and-so who did not know how to mount a horse. He soon calmed down, and we went off on a day-long inspection of part of the large ranch, which went quite well. I visited the mine quite often over the next few months, as was always my plan when I took over a new mine or appointed a new manager, to be sure that the operation was running the way I thought best. During every visit we set aside one or two days to visit the ranch. We had been out eight or nine times without incident and were on our way back one afternoon when the manager, who was following me, said that he really felt badly about losing his temper during our first trip to the ranch, but he had really never thought that he would see a man who had worked in so many parts of the world who did not know how to ride. He went on to say that he thought that I had improved dramatically and was riding very well and handling the horse in a good manner. Just then I was entering a deep, narrow trench in the trail, and a big snake came down one side. My horse was trying to get away from the snake and to get rid of me as well, but after the manager's praise for my riding I was determined to stay with the horse. It was a wild ride but I managed to hang on, although I developed a hernia as a result of my efforts.

On one occasion two of the staff members at Frontino were kidnapped. The government had decreed that people should not pay ransom because it only encouraged the kidnappers to go after others. I felt the same way, but after three months of listening to two wives begging us to rescue their husbands I relented. A ransom of 250,000 dollars was negotiated by intermediaries, and the company turned over the money. It was intercepted by the army, so another 250,000 dollars was paid, which did get to the kidnappers. However they were being pursued by an army group and were moving constantly. Our men were told that if the army caught up to them, they were to lie down and wave a white flag. This finally happened almost a month later, and the two men were taken out by the army even though their handkerchiefs, which they were using as a white flag, were far from white after several months in the jungle. They had been five months in captivity, moving constantly and often with little food. On Thanksgiving they had caught a wild turkey, and that is all they had for 12 men. Our men were both very thin, and one of them never quite recovered from the trauma. The company later recovered the 250,000 dollars that the army had confiscated.

Two of our dredgemen were later kidnapped at the South American Placers operation in Bolivia. Several well-armed men came into the camp and took the men off to the jungle. They were demanding that the government free some particular prisoners. Fortunately, the wife of one of the kidnapped men was a relative of the president, and she was able to get him to release two prisoners and send them to Chile. A few days later our men appeared in camp and soldiers were there to interrogate them. The soldiers then went off in the jungle to find the kidnappers. Three days later they returned with three men, whom our men identified as their kidnappers. The men were put in the back of a truck with army men guarding them and they left for Caranavi where the closest jail was located, about three hours away. However, when they arrived at Caranavi the kidnappers were all dead, shot in the back, and the army men said that they had been trying to flee.

We had one other kidnapping incident when the manager in La Paz was driving alone and was stopped and held at gunpoint in a small village he was driving through on the way from the mine to La Paz. The villagers demanded a tractor and a typewriter. After two days of negotiations the company turned over an old tractor and an old typewriter to the villagers, and our man was released. He told me afterwards that he was somewhat insulted that he was ransomed for an old tractor and an old typewriter, but I told him that he was an old manager so he should not feel bad.

Kidnapping was a very big business in Colombia during those years, with estimates as high as 80 million dollars per year. Many businessmen I knew were kidnapped and paid some very large ransoms. Some of them were killed even after ransoms were paid, and most men of means had bodyguards. I was walking two or three blocks to dinner one evening in Bogota with Bill Wilde when I commented that I felt as though we were being followed. Bill said that we had better be followed as he was paying the man, one of his bodyguards, to do so. Bill and I have been very good friends ever since he went to Bogota almost 50 years ago. He is the most successful entrepreneur I have known. I always traveled in a suit, but one time I arrived at our apartment in Bogota in old work clothes to be greeted by my very surprised wife, Sandra. I told her that the security men at the Frontino mine had learned of plans to kidnap me and had met me as I came out of the mine and that we had left by back roads without stopping for me to change clothes. We had been getting quite concerned about our children when we were in Bogota and would only let them go from our apartment to the country club by car, so we decided that it was not safe to take them to Colombia anymore. Their last trip was in 1976, although Kevin went with me later as an adult. I had other kidnap threats in Mexico and even one in New Canaan, so I was very fortunate not to have been caught.

For many years the only way to get to the Frontino Gold Mines was to go by ship from England to Barranquilla on the north coast of Colombia, then by riverboat for a week to Zaragosa, then eight days by mule to the mine at Segovia. When the company acquired a new hydroelectric plant from a British firm, an electrician from Scotland came out to do the installation. After the plant was operating well, the company offered the electrician a permanent job, which he accepted. He then sent for his fiancée in Scotland, who traveled out by steamer to Barranquilla then by riverboat and mule train to the mine. It was a long, slow trip. No one on the riverboat or the mule train spoke English, so the lady acquired a fair knowledge of Spanish on the trip. The only problem was that much of the Spanish, especially that of the mule skinners, contained more swear words than one would hear in the camp, so the lady swore like a mule skinner in all her conversations and developed quite a reputation for her foul mouth not only in Spanish but also in English. When I first took Sandra to the mine, I warned her not to be too shocked at the lady's language, but to my surprise and that of everyone else in the camp she never ever said a foul word in front of Sandra.

Labor negotiations were always very difficult at Frontino, and I would often be at the mine to support and help the manager. The union lawyer was a member of Congress and as such was authorized to carry a gun. One time after many days of negotiations we had settled most of the points but were still well apart on a wage increase, when we met after dinner to try to arrive at an agreement. As soon as we were seated, the lawyer put his gun on the table in front of himself and said, "Tonight we are going to settle these negotiations and do it on my terms." It was very tense, but the manager and I told him that under no circumstances would we, or our successors if he killed us, ever sign an agreement that the company could not live with economically. We argued until late without any progress. Shortly afterwards the head of security said that he had heard rumors of threats against my life, so I left camp in the middle of the night. Shortly after I left camp a dynamite charge was set off at the guest house, which destroyed the porch and part of the wall of the room where I usually stayed. The union negotiators disclaimed any knowledge of the attack on the guest house and the manager advised them that he was only authorized to settle on the terms we had proposed. They signed, as they remembered that after a long, bitter, and costly strike for them three years previously they had had to settle on the final company proposal.

Sandra loved to ride, and when we were at the mine one of the cowboys would bring up a horse for her. I worried about her going out riding, but one of the foremen who usually went with her was a tough-looking fellow with three notches on his revolver for bandits that he had killed, so I thought that she was well guarded. However, kidnapping became so rampant, and it was such a multimillion dollar business, that we decided that it was not wise for Sandra or our children to go to Colombia anymore. It had been such a nice country, and while we enjoyed visiting there for many years the guerrillas, the kidnappers, and the drug business destroyed much of what had been such a good country to work in.

The bull ring in Bogota was close to the Tequendama Hotel, where I usually stayed, and one Sunday afternoon I went to see a bullfight. I was wearing a suit and tie, my usual uniform when traveling on business, and on the way in I bumped into Dr. Rafael Samper, a well-known local physician who was assisting the mining companies in our efforts to improve the lives of the workers. He was dressed in very old clothes and an old, worn hat, and he said that the way I was dressed I would never be able to get out after the bullfight with anything in my pockets, as the thieves could strip a person clean in the crowded exits. So when I went out I had my arms crossed clutching the pocket with my wallet in one hand and the pocket with my passport in the other. When I got out of the arena the small bills and change I had in one pants pocket and a small knife and a handkerchief in another were gone, but I still had my wallet and passport thanks to Rafael's cautioning me. As I was walking past the parking lot I saw Rafael looking somewhat distressed and went to thank him for warning me, and he said that the only thing he had had in his pockets were his car keys and that they had been stolen. He was glad that I had escaped with the important items and said that the next time he would pay attention to advice that he gave to others.

One evening quite a few years ago, when I was taking the elevator up to my room in the Tequendama Hotel in Bogota, I noticed that a lady in the elevator had on a large ivory bracelet with a dog team on it. I said to her, "You have apparently been in Alaska," and she said that indeed she had and that her husband had been governor of Alaska and was now a senator in Washington, D.C. I introduced myself and told Mrs. Gruening that I knew her husband and hoped that I might be able to see him. She asked for my room number and said that she would give it to her husband, who was out at a business dinner. I worked until late and finally went to bed after midnight. Sometime after 1:00 a.m. there was a knock on my door, and it was Senator Gruening in his shirtsleeves and red suspenders. We talked at length about the guerillas and the developing drug and other problems in Colombia, and he was interested in hearing facts firsthand from someone who was directly involved in the local situation and had a different and more realistic view than some of the politicians and embassy people he had been talking to. Before he left we talked about Alaska, and he said that he remembered speaking to a group in Fairbanks during the celebration of the 50th anniversary of the gold rush where I was master of ceremonies as president of the Pioneers. I said that I remembered the evening very well and that we had apparently used the same book of jokes in preparing for the evening, as he used several of the jokes that I had been planning to use when I introduced some of the other speakers. He laughed and said that was the reason he always asked to be placed early on the agenda when there were several speakers.

The companies in Colombia were fortunate in having the services of an outstanding lawyer, Dr. Juan M. Arbelaez. He spent his entire career with the mining companies, and he was by far the most knowledgeable mining lawyer in Colombia and also did outstanding work on contracts for us in Bolivia and Peru. He knew enough English to read and converse on business matters, and when I first went to Colombia in 1953 without any knowledge of Spanish he helped me out speaking English. However, after a few months he said that it was necessary for me to learn Spanish and would not speak English with me any more. I really had to work at learning it and it was difficult for quite awhile. He was the company representative in Bogota, and after I was put in charge of the companies I spent a lot of time with him on government relations as well as on operating and management problems. He was quite reserved in our discussions during the day, but at five o'clock we would go to the Scotch Bar and after two or three drinks he would loosen up and tell me with increasing force, as the drinks flowed, what I should be doing in all the activities. We usually had quite a few drinks, then dinner, and talked until very late. Fortunately I was in the habit of taking notes, so even though I sometimes had trouble deciphering my notes the next morning I would have all his ideas and advice in hand. He was invaluable to me and we became very good friends, although never on a first-name basis: he was always Dr. Arbelaez to me and I was Mr. O'Neill to him. He was very sedentary, but I finally talked him into getting some exercise one time and for awhile he would walk the last few blocks to the office, but then a good friend our age, Dario Galindo, dropped dead on the golf course and that was the end of his exercising. We kept in touch until his recent death at age 91.

We had formed an exploration organization and learned of a potential dredging property in Bolivia that was in the national reserves. Manuel (Toti) Granier was on the planning commission in La Paz and knew of

Lew Harder. Toti stopped by the office in New York to speak about Bolivia's interest in obtaining foreign investment, so we mentioned the property our engineers knew about. Very soon thereafter, South American Gold and Platinum Company was invited to bid on the property. Prospect drilling crews were sent in, and good values were soon indicated. With Rene Rojas, an outstanding Bolivian lawyer, representing South American Gold and Platinum Company, negotiations with the planning commission in La Paz were started on a contract of exploitation, which went well until the problem of payment to the government hit an impasse. We had proposed a royalty basis with a percentage of production, which is normally used in mining leases, but the government insisted on a 50-50 split of the profits. We finally agreed to this, but then President Victor Paz Estenssoro said that he didn't even trust his own accountants and certainly would not trust ours. We had to work out a sliding scale royalty on production, based on values per cubic meter dredged. Without any previous mining experience in the area, however, it was very difficult to estimate the volumes that could be dredged as a basis for royalty figures. An agreeable formula was finally

With President Paz Estenssoro in LaPaz, Bolivia, 1963. Left to right: the president, Patrick, Rene Rojas, and Manuel Granier.

worked out, and a contract was approved by the Congress, giving the contract the status of a law. South American Placers was organized as the operating company, and I was elected president.

While we were in the exploration phase, we rented a small place facing the plaza in the village of Guanay. We slept with mosquito netting for protection against mosquitoes as well as bats, but one morning I woke up with an incision on my hand, which had been outside the netting, and it turned out to be a vampire bat bite. When I arrived in La Paz, I was tested for rabies and fortunately it was negative.

While construction was underway at Caranavi, a small village at the end of road transport, I visited on an Easter Sunday. A Franciscan priest, Father Venant, had left word a week before that he was visiting several villages in the area; he planned to say an early mass at a small village but would be in Caranavi for the Easter service. At about 11:00 a.m. the priest came out of the jungle on a mule, dirty and unshaven, and said that he wanted to clean up before mass. I told him that there were many of us who wanted to go to confession as well, so we put up a curtain in the door to the bathroom and he heard confessions while he was showering, shaving, and dressing. We put some boards on top of two oil drums for an altar, and I was the altar boy. It was different, but we were all happy to be able to attend mass on Easter Sunday.

A dredge that had been designed for airlift and flown into the interior of Papua New Guinea in 1938 was purchased, dismantled, and trucked to a port on a road that the Japanese had built during the war. It was shipped to Los Angeles for transshipment to Chile, where it was lightered ashore on barges at Arica. It was then moved by rail to La Paz, Bolivia, where it was transported by truck to the end of transport at Caranavi. Two trimotored Northrop freight planes that had been built for service in Korea but never used there were acquired. We built airstrips at Caranavi and another at Teoponte about 40 miles downriver, where the dredge was to be assembled. A camp including staff and worker housing, shops, a warehouse and power plant, offices, and so forth was constructed at Teoponte. A small church was also constructed later on, built mostly by the employees with material provided by the company. In all over 6,000 tons of machinery and equipment were transported in the two planes, and it was quite surprising that with the long move and transshipments only one small steel plate was lost.

Rubber tanks, each holding 1,000 gallons of diesel fuel, were floated down the river from Caranavi to Teoponte to supply the power plant. A man on a balsa raft guided each tank down the river, and the empty tank and the balsa raft were then flown back to Caranavi. This system worked very well for over four years, at which time the road was extended and the river transport and air service were discontinued.

The dredge was reassembled in about six months. At the inauguration of the dredge operation the natives insisted on sacrificing a cow on the deck, which was done before having a blessing by a priest. The president of Bolivia, Victor Paz Estenssoro, was there with several of his staff, and the operation got off to a good start. On one visit by the president I went to the guest house to take him to breakfast and was talking to his aide while we were waiting. When the president came out, the aide asked him if he knew that I had been a pilot during the Second World War. The president said, "No, but what did you do in the First World War, Mr. O'Neill?" I told him that I might look that old, but I really was not. Paz Estenssoro was president of Bolivia when we were negotiating the contract for mining in that country. At one point he visited New York, and we had dinner one evening at Club 21. He spoke very good English, and we had a very pleasant evening. Not long after that meeting, however, I went to see him at the palace in La Paz, and when I greeted him in English he responded in Spanish, saying that he would not speak English in the palace. I had to struggle with my limited Spanish whenever I met with him on business over the years; he would not even help me out when I was having difficulty getting a point across.

The government assigned inspectors to the operation, which we expected, but none of them had any experience in dredge operations. This led to arguments about the measurement and calculation of volume dredged, which continued through most of the life of the operations. The inspectors wanted to be involved in every aspect of the operation, especially the weekly cleanup when the gold was recovered. The company security men caught one of the inspectors stealing gold during one cleanup, and the head of the inspectors fired him at our insistence. Three years later, the same individual was rehired by the government, this time as head of inspectors, and he was very difficult to deal with. One cleanup day when most of the men in the camp were at the dredge this chief inspector arrived in camp and demanded to be taken to the dredge immediately. The accountant told him that everyone except himself was at the dredge, as were all the vehicles except for one old jeep, and that he was not a very good driver. The inspector insisted, however, and he and the accountant set off for the dredge. The inspector had one leg over the side of the jeep when it tipped over going too fast on a curve. His lower leg was badly smashed and had to be amputated below the knee. The company's insurance covered the expenses of the operation as well as flying him to New York for a prosthesis, and gave him 50,000 dollars for the loss of part of a limb. The government employees were very poorly paid, so this was quite a windfall. The man bought a vehicle and a house and was enjoying a better life and was much easier to get along with. About three years later this same man asked me if the company still had the same insurance. I said, "Yes, but surely you are not thinking about having another accident?" He said no, not really, but the way he seemed so embarrassed, I felt sure he had been considering the possibility.

There were Franciscan priests stationed at Guanay, a few miles upriver from the company camp. They were men who had served in the U.S. Army during World War II, then afterwards entered a seminary and were ordained. They were dedicated and capable. One of them, known as Padre Tex, was an exceptionally good mechanic. The first time I met him he was working on an engine in a boat at the camp. He had on old, well-worn hiking boots, dirty khaki trousers, a dirty T-shirt, and an old baseball cap, and he badly needed a shave. When they introduced him as Padre Tex, I wasn't sure whether he was truly a priest, since many people in the company called me padre. I was called padre because a company I was in charge of in Colombia had built a large church, and I had arranged for the company in Bolivia to provide materials and assistance so that the employees could build a church in Teoponte. The next morning at Sunday mass Padre Tex still had on the old boots and dirty khakis under his vestments, and still needed a shave, but he looked and preached like a saint. For quite a period of time Padre Tex was stationed at Mapiri, several miles upriver above the mining camp at Teoponte. When equipment broke down and the company employees could not fix it, they would call Tex on the radio. He would get on a balsa raft, and after several hours in turbulent rapids he would show up at camp and get to work. He could usually fix almost anything, and if he couldn't they would throw the equipment away and order a replacement. In return for his help Tex would pick out tools from the warehouse and shop that he needed, and they would put his balsa raft and several cases of beer in the plane with him and fly him back to Mapiri. Over the years Tex built many roads between villages, as well as schools, water systems, and so forth. One time I said that I wondered if he might not be more interested in building things than in converting people. He replied that he always insisted on men from the villages working one month a year with him, and they would start off the day with mass and end it with the rosary, followed by a lecture in the evening. He said that he got a lot of roads and other facilities built and made Christians out of a lot of heathens at the same time, which was true. During one period Tex was stationed at Guanay. Like most villages in that part of the world, the church was on one side of the village square and the town hall was on the opposite side with cantinas, shops, and houses all around, and the center area was used for soccer and other gatherings. It was customary for the priest to walk out in front of the church after mass to talk to the people. For a period of time the town leaders were communists, and they would walk across the square and heckle the priest. Tex was a big, husky man, and he ignored the taunts for quite awhile, but the men got meaner and meaner in their remarks. Finally one Sunday Tex had had enough, and he walked over and knocked each of the three men out. I was in Guanay not long after, and when I saw Tex I said that I understood he would just as soon hit some men as hear their confession. He replied, "Yes, and with some despicable characters, it is a lot more effective."

Victor Paz Estenssoro became president of Bolivia after a revolution in 1952 when the old tin barons and their associates were exiled or killed. On the road from La Paz down to Coroico and the lower country on the Kaka River where we installed a dredge the road goes up from 10,000 feet at La Paz through a pass at 16,500 feet, where it starts a steep descent to the mine area at about 500 feet above sea level. The road is literally carved out of the mountainside in many places, and there are sheer drop-offs of several thousand feet. At one curve in the road that had a sheer drop of 4,000 or more feet it was well known as the place to dispose of political prisoners: they would be shoved alive off the cliff to plunge to their death. One commentator once said that the drop was so great that they probably starved to death before they hit. The new constitution provided for a single four-year term for the president, so Paz Estenssoro stepped aside in favor of Siles Zuaso for four years and then came back in as president again. The miners' union in Bolivia was very strong, and attempts to modernize the mines or improve productivity were always thwarted by the miners. For example, the miners worked on production based on tonnage. When the grade of the ore dropped below economic levels in a particular stope, the manager would order them to abandon that stope and move elsewhere, but the miners would refuse to move because they were getting good tonnage, so the manager would fire them. Then the entire mine would go on strike until the manager was changed. Settling the strike always involved payment of wages for the time out on strike, so other mines would go on strike to get the same payment for no work that the first mine had received in its settlement. The mines were the principal source of the country's wealth, and they were operating at a loss. In previous years, whenever the miners did not get their way after a prolonged strike, truckloads of them, heavily armed, would drive in to La Paz to the Plaza Murillo in front of the presidential palace, and on more than one occasion the president or other officials had been hanged on a light pole in the plaza. One time toward the end of his second term I was standing with Paz Estenssoro at the window looking out over the plaza, and he said that it had been his intention during his second term to get the mines under control and profitable, but when the miners went on strike and truckloads of armed miners filled the plaza, he would look out at the throngs and the lightpost where some of his predecessors had ended their lives and he just didn't have the courage to hold out longer. He resolved that some other way had to be found to solve the problems in the mines.

Near the end of his second term, Paz Estenssoro was trying to change the constitution so that he could succeed himself, but there was much opposition. I was on my way from the mine to La Paz to have a meeting with the U.S. ambassador and President Paz the next day. During the trip I saw much more military activity than usual, especially as I entered the capital. There were rumors of a coup, but I went to the office the next morning and got in touch with the U.S. embassy. I was told that despite the rumors they did not anticipate any immediate problem, and I was to stop by the embassy about 9:30 a.m. to go with the ambassador to the scheduled 10 o'clock meeting with the president. Just as I was leaving the office, though, the embassy called to say that Paz was being overthrown and that I should go to my hotel as fast as possible and stay out of sight. The office manager gave me a small radio, and the chauffeur and I started off in a truck. There was a lot of shooting going on, and people were tearing up the cobblestone streets and erecting barricades. We plowed through and bounced over several small piles where they were just getting started, but then we came to one barricade that was about three feet high and getting higher. The chauffeur said that we could not go on, but I had just heard someone say on the radio that one of the mistakes of the 1952 revolution was that they had not killed all the foreigners, but they would do so now. I told the chauffeur that we had to go on, or I would be killed, and to blow his horn and go as fast as he could. We plowed through the barricade and bounced over the holes they had excavated and somehow made it to my hotel. My room was on the sixth floor facing the street, and the windows were about three feet from the floor, so I sat on the floor and peeked out the window occasionally. Across from the hotel was a hill where there was a military base. There were different factions in the military and the Air Force with its old World War II P-51 airplanes was strafing the installations on the hill right across from the hotel. There was a lot of shooting going on, and one stray bullet came through the window above and to my right. The hotel management sent word that meals would only be served during daylight hours in the back of the dining room and to use only the stairs. There was continuing coverage on the radio, and often they said that foreigners were not to be allowed to leave the country and should be killed. Except for going up to the restaurant during daylight, I spent the days and nights in my room listening to the radio and the frequent noise of shooting that went on all night. Often I could see the tracer bullets. It was a most uncomfortable time worrying about what was going to happen and how I would be able to leave the country. Early one morning it was announced that President Paz had been exiled and that the military was taking over. There apparently were problems deciding which faction of the military would name the new president, and there continued to be much activity and frequent shooting while I sat in my room except for mealtimes. Late one evening it was announced that an army general would assume the presidency and that law and order would soon be reestablished. The third day I was able to contact our lawyer to arrange safe passage for me to leave the country. Air service was restored on the fifth day, and I was able to leave although I had very nervous times at several checkpoints on the road to the airport when the guards said that they had not received any authority to let me go. However, they finally did let me go on.

The upper Tipuani River area, which was on one of the upper tributaries of the Kaka River, has been a well-known gold-producing area for many, many years. Most of the mining was done in open-cut hydraulic operations or shoveling into sluice boxes on the side hills, but there were also many miners who sank shafts on the side of the river and with pumps drifted underground to extract the gold-bearing material. It was dangerous work because the mines were often flooded, but it was a good livelihood for many men who worked in cooperatives. The village of Tipuani is on the side of the valley in a large bowl-shaped area; in the lower part the river goes through a narrow stretch between two large hills. The airplanes have to fly as low as possible in the V-shaped area between the two hills in order to land in the small bowl-shaped area. Other than mule or llama trails from La Paz the only access into Tipuani until very recent years was by airplane. There were hundreds of miners working in the Tipuani area, and it was well known as a lawless area. The miners worked in small groups as cooperatives, which were loosely affiliated under the leadership of a man known as Danny, who had a well-armed group known as Danny's 40 thieves. The company's concession covered the lower Tipuani River and extended up into the bowl where the village and airport were located. We thought there might be enough potential dredgeable ground in the bowl to justify installing a small dredge, so we sent a small crew with a prospect drill to evaluate the ground. Our crew had hardly gotten started before they were ordered out of the area by Danny and his group. The government had guaranteed peaceful possession of our concession area, so we protested. The president assured us he would resolve the problem, and a few days later a lieutenant with about 30 soldiers flew into Tipuani in a DC-3. According to our informant, after landing the soldiers assembled and just then Danny and his 40 thieves, who were better armed than the soldiers, came marching toward the plane. The lieutenant said, "Well, it looks like everything is well under control here," and ordered his troops back into the plane, which immediately took off back to La Paz. Despite frequent meetings and protests the government would not take any further action, and we never could obtain peaceful possession of that part of our concession area.

We had some difficult times getting the military men who took over the government to honor the terms of our contract. They finally agreed to do so but continued to give us problems on the interpretation of some clauses in the contract. Some of the officials insisted on being put on the payroll. There was no doubt that their salaries were very low, but it seemed to me there would be no end if we started paying one or two, apart from the fact that it was wrong anyway. As I mentioned earlier one minister of mines, at a party when he had been drinking heavily, even threatened me with a pistol demanding that he be put on the payroll. Fortunately, I was able to convince him that we would have to talk about it the next day and then told him that the parent company forbid me from making any such payments, and there wasn't any way I could hide such a thing from the auditors. To my surprise a few years later I found out that our office manager in La Paz had been giving the top labor leader 25 dollars per month from petty cash to avoid unionization of our workers. About 12 years after Paz was overthrown the country went back to open elections, and he returned to the country and was elected to one more term as president before retiring from politics.

When I was in the Army Air Corps flying school at Randolph Field in Texas in early 1942, there were three officers from Bolivia receiving flying training. Two of them, Cortez and Arce, were in the same company I was, Company G. The third, Barrientos, was in a different company. When I became an upperclassman I was the cadet commander of Company G. About 17 years later when we started operations in Bolivia I ran into Cortez and Arce, who had left the Bolivian Air Force, acquired two old DC-3 aircraft, and were flying commercially around the country. We occasionally used them, and on two or three occasions they flew me to the camp at Teoponte, which was at an elevation of slightly over 500 feet. One time Arce suggested that I fly, and we took off from La Paz at 13,000 feet and flew through the pass at 16,500 feet. When we left the mountains, we were over solid overcast as far as could be seen. I knew that there were no radios or other aids to help us, but as we passed one peak Arce told me to start letting down at 1,000 feet per minute on a certain heading, and he would time me with his wristwatch and tell me when to turn to a different heading. I was worried because I had not flown much since the war and was having trouble holding the course and the descent steady; however, we kept on, although I knew from previous trips in the clear that we were getting down close to hills.

After five or six minutes I mentioned my concern, but Arce said that we were letting down into a valley and that if we did not break out within two or three minutes we would go back up. About a minute later we came into the clear and were in a valley all right, but as I looked off to my left I saw the wreckage of a plane on the hillside. Arce said that the pilot had been off in his timing and didn't want to discuss it further. I told him that I thought they were taking high risks, but he did not seem concerned. I did, however, take stopwatches on my next trip for them to strap on the control column so that they could be more accurate in their timing. Cortez was killed a year or two later when he was flying up the narrow canyon to go through the pass at 16,500 feet. He was short of the pass when an engine cut out; the canyon was too narrow to turn around and he did not have enough altitude to get through the pass, so he crashed. Three or four years later Arce was flying a load out of Caranavi in his DC-3. Someone had put kerosene instead of gasoline in his tanks, and shortly after takeoff one engine quit. He had to turn into the dead engine, and the plane flipped over and went into a dive. He did not have enough altitude to recover and crashed. I did not know the other pilot, Barrientos, very well in flying school, but he later became president of Bolivia during the time when the military leaders were running the country, and I visited with him several times on business. He invited me to his home one evening, and we had a good time talking about flying school and flying in Bolivia. He was killed while in office when he took off in a helicopter and hit some power lines.

Julio Jaregui, who was general manager of the Bolivian operation for several years, was the nicest, most gentlemanly man I ever knew. He was a very small man, and one of the tin barons, Patino, had taken a liking to him when he was quite young and had sent him to England for his education. From then on he dressed and acted like a proper English gentleman, always with an elegant suit, hat, gloves, and so forth. One time he was driving me down to the mine—about a six- or eight-hour trip—and was telling me, while chain smoking, that his wife and three daughters had been after him for several years to quit smoking. He had been unable to stop, even though he knew he should and had tried to several times. I told him that I had been a very heavy smoker and that after several unsuccessful attempts to quit I had finally decided that I had to do it, and did. If I could do it, I said, I knew that he could. He asked if I could help him, and I said, "Yes, throw the cigarette you are smoking and all the other cigarettes you have in the car or in your suitcase out the window and over the cliff and don't ever smoke again." He asked if that was an order, and I replied that indeed it was. He said OK, and he never smoked again. He and his family thanked me many times, and he lived until past 95. I spoke to one of his daughters by phone recently, and she said that they were sure that their father would not have lived so long if he had not stopped smoking, and they were always very grateful to me for convincing him to stop. It is a great satisfaction to feel that I have helped a few people out during my life.

The road down to the camp at Teoponte was very narrow for a large part of the distance and also treacherous because it was along a hillside with very steep drop-offs, as mentioned earlier. Occasionally there would be slides during heavy rains and the road would be blocked. One time I was going down with a taxi driver; when he picked me up at the hotel in La Paz, I asked him to wait while I got a sandwich, and I asked if he would like one as well. He refused and was anxious to go, so I left with one sandwich and an orange. After five hours of driving we encountered a roadblock and turned around to go back, but another slide had occurred, and we were stuck for two days before the road was cleared. The driver and I shared the sandwich and the orange and slept for two nights in the car. He said that he would never again be in such a hurry as to not accept a sandwich when offered. As I write about this incident over 30 years later I am thinking about the great progress made in communications over the years. If we had had cell phones at the time, I could have called the mine and had them send a car, and I could have walked or crawled across the slide to meet it. Lacking such communication, I did cross the slide in hopes of finding someone who had driven up to find the road closed, but all I found were tracks indicating that someone had been there and turned around, so I went back to the taxi to have some protection from the rain while we waited.

One time when Sandra was with me visiting the mining operations at Teoponte in Bolivia, the manager's wife held a tea for her. All 18 wives in the camp were there, each with her favorite dessert. Sandra took two or three desserts, but then the manager's wife whispered to her that all of the women had worked hard preparing their own special desserts and that she should take one of each, which Sandra did. However she was very sick when she arrived at the guest house, so I took her to see the doctor, who induced vomiting. She felt somewhat better after two hours of rest, but then we had to go to a dinner party with all the staff and their wives at the manager's house, which was customary when we visited the different mining camps. I don't know how she managed it, but Sandra was very pleasant with everyone all evening. She did collapse, though, when we returned to the guest house.

We knew that the dredgeable ground in Bolivia was very hard, as indicated in the prospecting work we had done, and we based our estimates on 200,000 cubic yards per month. This was less than half of what a comparable-sized dredge was doing in Colombia. We often fell below our estimates, so the property was not as profitable as expected. The company did recover the investment and was modestly profitable for many years but was nearly finished with the available ground when International Mining was taken over in 1977. The new owners did not want to mine, especially in Latin America, and I was directed to dispose of all the operations. There was a Bolivian mining company that had operated in tin dredging, so I contacted the principal owner, whom I had known for many years. I told him that although the available reserves were limited there was a good block of high-grade reserves adjoining and partly under the village of Guanay. As a foreign company we had not been able to make a deal with the village, but as a Bolivian he might be able to. International Mining sold the Bolivian company, South American Placers, to Comsur at a very modest figure and Goni, the head of Comsur, was able to negotiate a deal with the village and moved the dredge to it. Just at that time the price of gold rose dramatically to the highest prices ever, so Comsur did exceptionally well. Goni's very substantial earnings from the Guanay dredging provided capital for his campaign and ultimate election as president of Bolivia, in which capacity he served with distinction for a four-year period. After a four-year hiatus he was elected to the presidency again and was trying to get approval to export liquefied gas through a port in Chile, which would have been of great economic benefit to Bolivia. However, the leftists developed great opposition because Bolivia had lost a port to Chile during a war many years before. The opposition became so strong and violent that Goni was forced out of office and is now in exile in the United States.

International Mining's exploration group did considerable prospecting in Peru, mostly on the eastern side of the Andes chain of mountains.

There was a road over the mountains to a small village, Quincemil, on the Inambari River. The river was incised in deep gravel deposits, and there were some gold mines operated with hydraulic monitors up above the valley floor. The gravels all contained gold in varying amounts, and the natives lived primarily from panning on the playas in the dry season. During the wet seasons the high rivers would cut into the banks and carry gold along. For as many generations as anyone could recall the same family had the right to a playa, where they would work each dry season. They would place rocks to act as riffles to catch the gold, and after the floods they would pan the gravel around the rocks to recover it; they made a living just as their ancestors had done in the same place. The playas were, in a real sense, gold farms with a new crop each wet season. During the wet season the natives would live and farm in higher areas. I thought it especially interesting that without any legal documentation they all respected each family's rights to certain playas.

We were looking for dredging ground, so we drilled in the river valley but the depths were great and the values too low for dredging, so we moved several miles upriver. A small crew would be sent in on trails over the mountain with axes, shovels, and wheelbarrows to clear an area where a small plane could land. We used a Helio Courier that could operate out of very small areas, and the drill equipment and tents would be flown in. Where they had to cross the river, they would install a cable with a small board to sit on suspended from two wheels, and you would pull yourself across. The river was very wild so this was a quite a thrill. We did not have any success in developing a dredgeable area on the Inambari River, so we moved out to the Madre de Dios River, which was a big river in a very wide valley where we had small boats to establish camps and drill sites. One time, the engineer in charge and I spent several days going a long way up the river to look at some of the tributaries as potential prospecting areas. One afternoon we were several miles up a tributary that looked interesting, and we had just pulled ashore to look around and spend the night, when a group of wild-looking Indians came out of the jungle with their bows and arrows at the ready. They were painted up with sticks in their noses and big earrings and were wearing only small loincloths. They came toward us in a very menacing way. The engineer said, "On closer look this doesn't look like a very promising area to prospect, Mr. O'Neill." I readily agreed, and we jumped back into the boat and left. The exploration on the Madre de Dios was not successful, so we left the area and to this day, 35 years later, I have never heard whether anyone else ever did any prospecting in the area we so quickly left.

There were small, isolated communities in the jungles of Peru that were governed by locally elected men, and they had strict regulations that everyone in the community abided by. We had a camp adjoining one of these communities near where we were drilling, and one of our crew got fresh with one of the young girls there. The leaders had a public trial and sentenced our man to 10 lashes, which were administered immediately. Our drill foreman took the man back to camp and doctored him as well as he could, and we never had any more trouble in that area.

During the time we were exploring in Peru, one of the officers of International Mining invited me to his home in Bedford for dinner. There were other guests there as well, and during cocktails the host was telling one couple about our exploration in Peru and my strenuous trips in the jungles, and about how I had gone over the Andes with mules and llamas and down turbulent rivers on balsa rafts. The couple was quite impressed and commented on my difficult and hard work in the jungles. A month or so later, I was in Lima settling up some accounts with the widow of a pilot who had done considerable flying for us in Peru but who had crashed a few months previously ferrying in a new plane. The widow and I and our accountants finally settled everything late in the day, and as we were leaving I asked her if she would like to have dinner with me. She said that she had not been out since Gene had died and felt that it was about time and that there was a very nice restaurant out of town that she would like to visit. I picked her up later, and we went to the restaurant, which was indeed very nice. She was a very attractive lady, and we were having a cocktail sitting in front of a large fireplace in very elegant surroundings when in came the couple I had met in Bedford. They exclaimed, "Man, you really do lead a tough life don't you!"

The Fresnillo Company in Mexico was another company that International Mining was able to buy control of at a reasonable price from English owners who, as I have mentioned earlier, were suffering from double taxation. The original mine was named after the town, Fresnillo, and the company had developed several other mines in more recent years. The Spanish had produced silver from the Fresnillo mine in the mid-16th century, and there had been mining activity there ever since. The English company had been operating the mine very profitably for over 150 years. Numerous veins had been exploited down to depth, but it had been quite awhile since any new veins had been discovered. There had also been new taxes imposed over the years, so the outlook for the old Fresnillo mine was not good. In fact, taxes had been increased to the point where most of the mines in Mexico were unprofitable, and most were foreign owned. In 1961 the government issued a decree stating that mines that became at least 51 percent Mexican owned would be exempt from most of the onerous taxation, but a fair portion of the earnings that resulted had to be spent on increased exploration for and development of other mines. The first group to Mexicanize was Penoles, which had several mines and smelters; the Fresnillo group, which was controlled by International Mining, did so soon after. Earnings from the profitable mines in the group were partly used for exploration at the old Fresnillo mine, which appeared to be nearing the end of its very long life with very few reserves left. The directors authorized an extensive drilling program to the west of the old mine, which was thought to be the most likely area by the geologists. More than a year of drilling did not develop anything to the west, so a drilling program to the east was initiated. After more than a year of intensive work without finding any new veins the directors decided to abandon further drilling. The manager asked for permission to drill one more hole where the drill was being set up, which was granted. That hole hit a vein with high-grade ore, and when they surveyed the location they found that the surveyor who had staked out the hole had made a mistake and that the hole was 50 meters away from where it was supposed to be. Subsequent drilling and underground development revealed that if the hole had been drilled where it was supposed to be it would have missed the vein. As it turned out, further drilling revealed an extensive high-grade vein, and subsequent underground development found others as well. The old Fresnillo mine became the biggest and best producer of silver in the country and still has a long life ahead at the time of this writing, which is over 20 years since the new area was opened up.

Other mines were also developed after Mexicanization. One of the largest was at Guanajuato, which was a well-known mining area with a very long history of production. A Canadian geologist and professor, Bill Gross, whom we knew from other mining interests in Canada, came to see us in

Patrick with associates at Torres Mine, Guanajuato, Mexico.

New York. He had been doing extensive geologic work in the Guanajuato area over several years and was proposing a drilling program to test his theory about a faulted displaced section of a major vein that had not been mined. He had presented his ideas to companies in Mexico without success. His thorough research and geologic theories looked interesting to us, however, so we prevailed on the other directors of Compania Fresnillo, S.A. (the Mexicanized company) to take on the project. The first drill hole hit a vein, as the geologist had predicted, and subsequent drilling developed a substantial ore body. A shaft was sunk 2,000 feet, a mill was built with the necessary shops and supporting facilities, and a very important and profitable mine was put into operation, which is still in production almost 30 years later. When the mine was starting up, we held a meeting of the directors at Guanajuato, and most of us took our wives along. Several of the wives wanted to see the mine, so we all got outfitted with boots, coveralls, hard hats, and lights and got into the cage to descend into the mine. The foreman pressed the button signaling the operator to descend, but nothing happened as there was a power failure just at that moment. The power was off for three days due to a major failure at the generating plant; if we had descended before the power failure occurred, we would have been in very serious trouble as the only way out of the mine was to climb 2,000 feet on vertical ladders, an extremely difficult task for the ladies and an almost impossible task for one older, overweight director. I have often thought how very lucky we were that the power failed before we went down, and I get very upset when I think about what a tremendous task it would have been to get all the directors and wives out.

We had an exceptionally good working relationship with our Mexican partners, and I always enjoyed our regular bimonthly meetings and frequent visits to the different mines. The chief executive officer of Penoles was a very efficient and capable German. He had been a member of the elite SS Group in Germany at the time of World War II and had been in a small group that had tried to assassinate Hitler. He managed to escape when their mission failed and was able to get to Spain. While he was trying to get a permit to go to the United States, he met and married a U.S.-born secretary at the embassy, and so was able to go. He went to work as a metals trader in New York, was later transferred to smelter operations in Mexico, and ended up as CEO of Penoles in Mexico City. The board of Compania Fresnillo, being 51 percent owned by Penoles and 49 percent by the Fresnillo Company of New York, had five directors from the former and four from the latter. In order to expedite matters at the board meetings when there were problems with language and understanding of technical matters, an executive committee was established with four technical men from each side. We met early for a breakfast meeting at the University Club the days of the board meetings, and all the matters pertaining to the operations were settled at the executive committee meetings and then presented for board approval. It was a great working relationship; several new mines were developed, and the company was very profitable. The German CEO and his wife had two daughters. When he and his wife traveled without the children, which they did quite often on business, they would go on different planes for security for their children. Most unfortunately, when the older girl was in her late teens she developed a brain tumor and died. Three years later the other girl was driving home from college to their place in Texas and was killed in a head-on collision with a car that was passing a truck. Everyone who knew them was terribly distressed and heartsick; they were such a wonderful couple. They could hardly stand to see other children because it made them so sad, but after three or four years the wife invited Sandra and our children to lunch one time when we were in Mexico for meetings.

Sandra and our children, who were about ten and seven at the time, still speak at times about the wonderful luncheon with elegant table settings of silver and crystal and delicious hamburgers served on sterling silver trays by white-gloved, uniformed waiters.

After International Mining was taken over by Pacific Holding in 1977, the Fresnillo Company was merged into Rosario Resources. I joined Rosario as executive vice president and continued on with the group in Mexico; I also looked after Rosario's other mining operations in Honduras, Nicaragua, and the Dominican Republic. A few years later Rosario was taken over by AMAX, a very large mining company, and I continued on with them as senior vice president. However, after many years with smaller companies, where decisions about acquisitions or other matters could be made quickly, I found it very frustrating to lose opportunities while proposals were going through various committees and endless discussions, and I retired and started consulting. As a consultant, I continued with the Mexican operations for several years. The 24 years that I was on the board of Compania Fresnillo and their subsidiaries were very interesting and challenging times

At an exploration site in Mexico. Left to right: Patrick, Ricardo Chico, Luís Villasenor, and Pedro Sanchez Mejorado.

Patrick on a mule at an exploration site in Mexico.

working with excellent engineers and executives whom I admired very much. The board met every other month, and I often went down in other months to visit the mines or exploration projects. In 24 years on the boards, I only missed one board meeting when I had an operation. AMAX later sold their interest in Fresnillo to Penoles, which I never understood as the Fresnillo properties were very profitable and continue to be to this day, but AMAX itself was falling apart. Penoles is the largest silver producer in the world, mainly due to the Fresnillo mines. Quite a few years after I retired I was invited by Penoles management to visit the Fresnillo operations. My son Kevin went with me and we had a great trip. It was wonderful to be greeted so warmly by so many old friends whom I had known and worked with for so many years. It was gratifying to see the improvement and expansion in all aspects of the operations, as they had continued to develop more reserves and expanded all the facilities. Kevin and I went underground at the mines and visited all the surface plants and offices, and we were treated royally both at the mines and at headquarters in Mexico City. It was a gratifying feeling to know that I had played a part in the development of the Mexican operations for 24 years and to see such continuing development.

One of the men I traveled with to Mexico occasionally was Bob Hoffman, who had interests in another mining company in Mexico. Bob was Jewish, and his parents had emigrated to the United States, where his father was a tailor in Boston. Bob was the oldest of three brothers and often got into fights with Irish and other boys in school and on the streets. A Catholic priest who was teaching boxing in a gymnasium in the church basement saw him fighting and suggested that he come over to the church gymnasium, where he would teach him to be a better fighter. Bob did go to lessons and became a professional fighter. He worked his way through Harvard and also put his brothers through Harvard. He graduated as a geologist and went to Canada to seek his fortune. He discovered a mine there, as well as one in Minnesota and one in Mexico, and became a very wealthy man. He donated money for a new geology building at Harvard, which is called the Hoffman Building. In addition to making very substantial donations to Jewish charities he gave a million dollars to the Catholic Church in appreciation for the priest who taught him boxing, and then gave a million dollars to a Protestant Church so that people would not think him prejudiced. Despite being very wealthy, Bob was also very frugal. When the price of a New York shoe shine was raised to 25 cents in the early 1960s, he would wait until he went to Mexico to get his shoes shined for 10 cents. He always stayed in an inside room in an inexpensive hotel in Mexico City. One time while we were waiting for a delayed flight at the airport in Mexico, Bob asked me if the shirt I had on was a wash-and-wear fabric. I said it was, and then he asked if I washed it myself, which I assured him I did. Then he asked, "Well, why does your shirt look so much better than mine?" I told him that it was because I carried a small travel iron and pressed my shirts before wearing them. He then said, "My poor Jewish tailor father must be turning over in his grave to hear his son learning something about tailoring from a damned Irishman."

Although they lived in Manhattan, Bob Hoffman and his wife had an old farm in Connecticut with an old house and barn and many acres of abandoned farmland. Bob started planting trees and was up to a million trees before he died. He spent a large amount of money rebuilding the old barn and installed very nice cabinets and drawers for his maps and mineral specimens. However, his wife could not get him to spend anything on the old farmhouse, which had a hand pump for well water and an outhouse. The first time that Sandra and I saw their place in Connecticut we were invited out for dinner on a Sunday. When we arrived, we realized how overdressed we were, especially when Bob said that we were going to a "lobster in the rough" place on the shore at Noank, where you eat on picnic tables. Sandra, being a southerner, loves crab, lobster, and crawfish and really knows how to eat them. Bob was very impressed and told Sandra that she was his kind of woman, and he was a great admirer of hers from then on. Bob had diabetes, as do our two children, so we had much in common. He was quite helpful when I was raising money for the Joslin Diabetes Center in Boston. He did not take very good care of himself, however, and he died from complications of diabetes after having one foot amputated and later a leg. His widow soon sold the property in Connecticut, obtained a much bigger apartment in Manhattan, and lived luxuriously for the rest of her years.

There was a very large pile of mill tailings alongside the town of Fresnillo, which was a nuisance for the townspeople when it was windy and the fine sand blew into their homes. Seeding of tailing piles was just getting started in the early 1960s, and I suggested at a Fresnillo board meeting that consideration be given to doing it at Fresnillo. One of the engineers was sent to a conference in Tucson on the subject and to see mines where it was being done. The seeding was very successful, and it was not long until grass and bushes covered the tailings, resulting in greatly improved relations with the town. The seeding program was extended to all the mines in the Fresnillo group. It was not very costly to do, and it was very gratifying so see green hills instead of barren piles of sand.

Fresnillo had a large very successful mine, Naica, in Chihuahua in northern Mexico. It was a very wet mine with large volumes of very hot water, which was pumped to the surface and then used for irrigation in the arid area around the mine. Even with extensive ventilation and blowing refrigerated air into the mine the heat created difficult working conditions. It was so humid that it was very enervating. The workers obviously got used to it, but I would be exhausted when I came out after four or five hours climbing through the stopes. It was a good camp with comfortable houses, a nice clubhouse, bowling alleys, a swimming pool, tennis courts, and a nine-hole golf course, but it was a long way from any village or town. They even had some horses, which Sandra loved to ride whenever she was there with me. One day after we came out of the mine the general manager and the local manager and I drove out to see the golf course, and we saw two people galloping towards us. The general manager, Luis, said, "There are some real cowboys: look at them ride." We soon saw that it was Sandra and a cowboy, and the men were amazed at the way she was riding. She always liked going to the mines with me, and the people in the isolated camps always enjoyed having her visit since she could speak Spanish and often brought things to them from the States. One time when Sandra and I were visiting one of the mines with Luis we were on an upper level on a dump and wanted to go down the dump to a lower level, but Sandra had high heels on, so Luis gave her his boots and we slid down the dump while Luis drove down with her high heels.

Rosario had a small mine at Huautla in Mexico, and one of the 51 percent partners was Juan Gallardo, a tall, handsome man. When Juan was in college, his wealthy father had insisted that he work during vacations, and one of his jobs was as a busboy at a fashionable restaurant in Mexico City. The headwaiter liked Juan and taught him to be a good waiter without knowing anything about his background. Two or three years later Juan became engaged to the daughter of another very well-to-do family, and her father had a big dinner announcing the engagement at the restaurant where Juan had formerly worked. When the headwaiter approached the head table, he saw Juan and exclaimed, "My God, Juan, how did you ever make it?" It was embarrassing at the time, but Juan did marry the girl, and often enjoyed telling the story afterward.

International Mining got involved in chrome ore in Turkey for a few years. John Caouki and his family firm, Egemetal Madencilik of Istanbul, were exporting in bulk chrome ore that had been hand-sorted to upgrade the material. They had several mines that were also producing ore that was not of a high enough grade for bulk shipping, and they needed capital to build a concentrating plant. The board of International Mining approved an investment of two million dollars, and I went over to arrange the project. Sandra and I were on a trip to Europe at the time, so we visited Egemetal's marketing and export office in Geneva and left our daughter and Sandra's mother in Geneva while we went to Turkey. We flew to Istanbul, where we found that John Caouki had suffered a heart attack in Iskenderun in the south central part of the country where the ores were shipped out. We flew with John, Jr. to Ankara, then drove eight or nine hours to Iskenderun, arriving after midnight. We rented a room in a small, very old hotel that smelled so strongly of urine we could hardly stand it, but we opened the windows wide and went to bed. About 5:00 a.m. we were practically ejected from bed by the call to prayers from a loudspeaker in the minaret right outside our room. What a shock! After our rude awakening we went to the hospital to see John, Sr. and found him out of danger but confined to bed for ten days or more. We visited the port to see where they consolidated the ore shipments and had them sampled, assayed, and prepared for export and spent some time reviewing their assay procedures. We also spent a day sightseeing, including a trip toward Syria where we visited some of the old caves where the early Christians had gathered south of Antakya. Then we went to Bursa, which is across the sea of Marmora from Istanbul, and from there drove to Orhaneli, where the mines were located and where it was planned to build a concentrating plant. Along the road there were cherry trees and people selling cherries, so we stopped to buy some and bought a basketful for one dollar, including the basket. At the proposed plant site they had set up a picnic table and roasted a lamb for lunch. However, there weren't any toilet facilities there, so when Sandra needed them we went off in the car to the small village of Orhaneli. John, Jr. stopped and went into several places but came out shaking his head each time. Sandra finally said that she could not wait any longer, and he said that he was trying to find a Western-style toilet for her but that there weren't any in Orhaneli. All the toilets in that area are Turkish style, meaning that there is a place to put your feet on each side of a hole that you squat over. Sandra said she didn't care what it was: she could not wait any longer, so he stopped at a service station where they had a Turkish-style toilet.

After lunch we drove toward the mines and dropped Sandra off at the foreman's home, which was a small, rough, eight-by-ten-foot building where he lived with his wife and little daughter. The wife could not speak English and Sandra could not speak Turkish, but when we picked her up about four or five hours later she had learned an unbelievable amount about the woman and her family and her life in Turkey. The mines were small underground workings that produced ore that was hand-sorted by women as it was brought out of the mines. The higher grade was exported, and the lower grade was being stockpiled for processing by the proposed plant. We agreed on a site for the plant and then went on to Istanbul for organizational meetings. John Caouki and his three sons who worked with him were all fluent in seven or eight languages; John's wife, who was a descendant of the Jewish people who had been expelled from Spain many years before, preferred Spanish but also spoke Turkish, French, and English. The family and their associates spoke Turkish in business and French at home and socially. They were interesting people to do business with and always lots of fun socially. When we dedicated the plant, I gave the dedication speech in Spanish because more people there understood Spanish than English.

An engineering firm in Istanbul designed the concentrating plant, and we went ahead with the project. I soon learned I was to have a major problem with what they called bakhshish. For everything of any consequence that one bought or contracted for one was given a bill of sale for approximately 50 percent of the price; the other 50 percent was under the table and was known as bakhshish. It was an accepted way of avoiding taxes, but it was an impossible way of doing business for a foreign firm being audited by a major international auditing firm. The directors of International Mining understood when I had receipts for only one million dollars but had spent two million dollars in the construction of the plant, but the auditors gave me a really bad time. Our partners were quite irresponsible about maintaining quality in the shipments, and we had considerable litigation about the grade of the ore exported, so we operated in Turkey for only a few years before selling out to our partners.

John Caouki's sons were not the same caliber as their father, and after he died they often had problems delivering what they had contracted to do. I did keep in touch with them for years at Christmastime and other special occasions and saw them whenever they came to New York, but then one year the cards were returned and the telephones were disconnected, so I did not know what had happened to them. About three years later while shopping with Sandra in Las Vegas, I was sitting near the dressing room while Sandra was trying on some clothes, and there was a gentleman sitting next to me whose wife was doing the same thing. It sounded to me like they were speaking Turkish, which they were, so I asked him if he knew anything of or about the Caoukis. He said that he had known the family for years and that they were in trouble with the government about taxes and had fled the country, and he had no idea where they were. I have never heard anything more about them and probably won't.

When International Mining acquired a one-third interest in Pato in 1954, Pato had an exploration program in Brazil on the Rio Jequitinhonha, where both gold and diamonds were present in the river gravels. It is very difficult to evaluate the value of diamonds in an alluvial area as just one diamond can give a good value to a large area, so a drilling program usually used on gold properties was done in the hopes that enough gold might be found to pay the cost of dredging, and that whatever diamonds were recovered would then be the profit. The project was supervised by an old hand in Brazil named Jack McCarthy, whose wife, Angie, was a sister of General MacArthur's wife. They were a very interesting couple, and I spent many enjoyable evenings at their home in Belo Horizonte listening to their trials and tribulations during many years of prospecting and mining for diamonds in Brazil. Jack had obtained the permits for prospecting—*lavras*, I believe they were called—but after three or more years of prospecting, the applications for permits to operate—*pesquizas* I believe they were called— were denied for all foreign-affiliated organizations by the new president, Goulart. We had to abandon the project, not too unhappily as not enough gold had been found to justify installation of a dredge just for gold. At the time our applications were denied we were investigating drilling much larger holes, sinking caissons, or installing a pilot dredge; and later on another group did as we had planned. A Brazilian group along with Pacific Tin Consolidated and others later took over the property after a change in the mining code. I was a director of Pacific Tin so was once again involved. This group installed a small pilot dredge, which confirmed that there were enough diamonds with the gold for an economical dredging property. Two dredges were installed and operated for over 20 years. I had previously visited a gold dredge operated by Brazilians and was surprised at the lack of concern about security against stealing, as theft from gold operations is well known all over the world. On the dredges in Colombia the recovery tables were all screened in with high-voltage wires over the tables, and the rooms where the concentrates were gathered were especially well protected. In Brazil, however, the gold recovery and concentrating areas were wide open. Things were even more open on the dredges recovering gold and diamonds. The concentrates with the diamonds were spread out on a table, and anyone who had spare time would sort through the concentrates picking out the diamonds and dropping them into a container. One time when Sandra was there with me she enjoyed picking out the diamonds and spent several hours doing so one afternoon. When we were preparing to fly back to Belo Horizonte late that afternoon, the dredge superintendent asked if we would take some diamonds with us to the manager in Belo Horizonte. Just as we were leaving, one of the Brazilian partners, George Reed, decided to go so he was handed the small sack of diamonds. We were at the same hotel and went together to dinner at the manager's home. While we were having cocktails, the manager asked if either of us had brought in some diamonds. George said that he had but that he had left them sitting on the dresser at his hotel room and would go back for them immediately. The manager said not to worry; he would take us back to the hotel after dinner and pick up the diamonds then. Somewhat to my surprise the bag was still on the dresser. Most of the diamonds produced in Brazil are industrial-quality, but there were some of gem quality in the bag as well, and the value was estimated at 50,000 dollars. The manager of the mine, Alexander Misk, was an old friend who got off to a great start in life with three big events in one day. He graduated from college, got married, and won the lottery all in the same day.

Pacific Tin Consolidated was an outgrowth of Yukon Gold that was formed by the Guggenheim group in the early 1900s to dredge for gold in the Klondike, and when the reserves there were finished they moved their dredge to Flat, Alaska, where it operated for several years, and then to Malaysia to dredge for tin. I was elected to the board of Pacific Tin in 1975 and was pleased to be affiliated with a Guggenheim company, as my grandfather had been working for a Guggenheim company in Alaska when he had sent tickets for my parents to travel to Cordova, where I was born. Pacific Tin later acquired the Feldspar Corporation in North Carolina, and several years later control of the company passed to a Canadian company and the name was changed to Zemex. Peter Lawson Johnston of the Guggenheim organization, who was chairman of Pacific Tin, continued on as chairman. After I passed retirement age the board passed a special resolution each year keeping me on, so I was 87 when the company was sold and I was fully retired. It was the first time since I was 15 years old that I was without an income of some sort, and it was a very traumatic experience.

During my first visit to Brazil in 1954 I had an opportunity to visit the famous San Juan del Rey mine about four miles southeast of Belo Horizonte. An English company, this mine had been operating continuously since 1834 as a major gold producer. For many years it was the deepest mine in the world, and because it was very hot it was the first to introduce refrigerated air into the mine. In the early years the mine was supplied entirely by mules bringing everything in more than 60 miles from the coast. They had over 1,000 mules hauling in supplies. All the equipment, pumps, hoists, compressors, and the power plant, were designed to be broken down for carrying by mules, singly or in tandem. The long wire cables that were necessary for mining at depth would be transported with a few loops of the cable on each side of a mule and then back to the same thing on the next mule, and there would be a string of 50 or more mules transporting one continuous cable.

The manager's house, which also had several guest rooms, was the most elegant I had ever seen at a mine. The toilets and sinks were English blue Spode, the beds were large four-posters with canopies, and there was a pull cord by the head of each bed that you pulled when you awakened so that a servant would bring your morning tea promptly. The furnishings were elegant, and there was a grand piano in the sitting room, all of which had been brought in overland by mules. The silverware, crystal, and dinner service were all English, and they even had special plates for fish. Both red and white wines were served with each meal by white-gloved, uniformed waiters carrying everything on sterling silver platters. It was hard to imagine such elegant service and wonderful food and wines in such an isolated area. The manager, who was planning to retire, had had that position for 34 years. Every Sunday the entire staff had to pass in review in their Sunday best on the lawn in front of the big house while a band played and the manager sat on the veranda petting his dog. When I went back to the mine almost 20 years later, most of the niceties were gone but the mine was still working, although not as profitably as in earlier years, according to what I was told. No one answered when you rang for tea, the wine cellar was quite limited, and the staff was smaller and not as well disciplined. The mine was deeper than when I had been there previously and hotter, even with the refrigerated air being circulated. The wall rocks were so hot that if you splashed a little water on the rocks tiny chips would fly off. A visit underground was like taking a prolonged steam bath.

In the early years of the San Juan del Rey mine, the company obtained property and mineral rights covering the valley upstream to the ridge top of the hills, to protect their water rights and for the timber that they needed. Large deposits of iron ore were found near the top of the hills in recent years. To obtain the iron ore an American iron company bought the English company. The iron company was not interested in the gold mine and considered closing it; however, the cost to close it would have been very high, over eight million dollars just for terminating the employees, and there would have been other closing costs as well. The iron company knew that Brazilian owners could liquidate the company and terminate the employees much more easily and at less cost than a foreign company, so they separated out the iron ore and sold the company to two wealthy Brazilians for two dollars. One of these Brazilians had recently acquired an interest in the Araxa columbium mine with Pato and Moly Corporation, another subsidiary of International Mining, so he asked me and Dan Kentro of Moly Corporation to advise him on what he should do with the mine. Dan and I spent a week or more at the mine reviewing in detail the operations and the geologic studies and reports, and then went to Rio to meet with the owners. We told them that if we did not own the mine, we would not buy it, but if

we did own it, we would not sell it. We went on to say that although production had been dropping in recent years and mineable reserves were at an
all-time low and the iron ore group thought that the mine should be shut
down, it had been my experience in Colombia and Mexico that these old
mines didn't die easily if given a decent chance to survive. There seemed to
be possibilities of finding more ore in the extensions of previously worked
veins and developing other veins indicated by exploratory drilling and old
workings in adjoining areas. We also suggested that they consider a joint
venture with a major mining company that had extensive experience in gold
mining. There were two English–South African companies exploring in
parts of Brazil at that time because the geologic conditions are similar to
parts of South Africa, and they were following up on the continental drift
theories. We suggested that they contact one of those companies, and they
made an agreement with Anglo-American, which undertook a very aggressive exploration program that developed substantial reserves. The company
became very successful again, and after a few years Anglo-American bought
out the Brazilians. The company is still operating, and gold production is
at high levels at the time of this writing, 25 or more years after the Anglo
group took over the operation.

Moly Corporation, a subsidiary of International Mining Corporation,
had for many years a partnership with Wah Chang in a Columbium mine
in Brazil. A very high-grade columbium ore body had been discovered, but
it was a very difficult ore to treat. Several years were spent on developing
a process for the beneficiation of the ore in order to produce a market-
able product. About the time the treatment problems were resolved Wah
Chang died, and his interests were sold to settle estate problems, which
were complicated by his having two wives. Part of his interest was bought
by Moly Corporation and part (one-sixth) was purchased by Pato Consolidated Gold Dredging Ltd., of which I was then chairman. The company
had serious tax and export problems that had to be resolved before the
company could export their product, which was essential for the success
of the enterprise. Just over 50 percent of the company was owned by a Mr.
Meloviani, who had not been able to resolve the tax and export problems. A
banker who had been finance minister and an ambassador with high-level
connections and an excellent lawyer on his staff agreed to purchase Mr.
Meloviani's interest if the company could convince him to sell.

Several attempts had been made, but Mr. Meloviani was not willing to sell. I was getting along quite well with people while operating mines in other Latin American countries so was asked to go to Rio de Janeiro to see if I could convince Mr. Meloviani to sell. Arrangements were made, and I went to Rio and showed up at his office at 11:00 a.m. on the agreed day. When I entered his office, I greeted him in English. He responded in Portuguese, saying that he did not speak English. In my mediocre Spanish, I asked if he spoke Spanish to which he replied no. So I suggested that if we both spoke slowly, he in Portuguese and I in Spanish, perhaps we could understand each other enough to negotiate. He placed a bottle of Johnny Walker scotch and two small glasses on the table between us, and we started talking, but had some difficulty understanding each other. By 1:00 p.m. the scotch was mostly gone, and we were not making much headway in our negotiations. He suggested we continue talking over lunch and we went to a very nice French restaurant nearby. We each had a martini then a bottle of wine with our lunch. By the time we finished lunch neither of us was in any condition to discuss business intelligently, so I said that I had another meeting and asked if we could continue our meeting that evening, to which he readily agreed. I went to my hotel to sleep and sober up, and I was sure that he did the same.

We met again that evening at the Copacabana Hotel, and after some discussions and two drinks we left for a very good restaurant two or three blocks away, where we hoped to be able to reach an agreement. Mr. Meloviani was considerably older than me, and as we walked along the street he held my arm. An attractive young woman approached us, and as she came abreast she spun around, grabbed my other arm, and said in English, "Don't fool around with that dirty old man; come with me, and I will show you a good time." He immediately said in excellent English, "Don't bother us, you little whore." She left and I said, "Damn you, here I have been struggling all day in Spanish and you speak English as well as I do." To which he responded, "Yes, and I speak Spanish a hell of a lot better than you do, but let's go along to dinner and we can talk." He said that he had been involved with the Columbian project from the start and really did not want to sell, but he had not been able to resolve the export problems and had decided that he must. He said that he had not yet made that decision when he agreed to meet with me and had stalled for time by being difficult with me and creating a language problem. He went on to say that he had enjoyed the day with me and would go ahead with the sale, which he did.

Rosario Resources had taken over some old mining operations in Nicaragua several years before I joined the company in 1978 as senior vice president in charge of their mining operations. When the Sandinistas overthrew the Somoza regime in 1979, they announced that they were going to expropriate or nationalize the mining operations, as well as many other industries in which the Somozas had interests. A lawyer and I obtained meetings with two of the top officials in the Sandinista group. We did not dispute the fact that the Somozas had interests in some of the mines, nor did we dispute their right to nationalize mines. We suggested, though, that since we had experienced men operating the mines and legitimately owned the machinery and equipment we should make a joint 50-50 venture. We went on to explain that the mines were only marginally profitable and the only way that the government would have any income from the mines would be to keep them operating as they were. The loss of discipline and efficiency by putting inexperienced revolutionaries in charge would very soon result in losses and a financial burden to the government. We negotiated with officials in the Banco Central for several days and were on the verge of signing an agreement when word came that another top man in the Sandinista group had gone out to the mines and announced expropriation and put some of his men in charge. Our manager and foreign staff were ordered to leave immediately. It did not take long for the new operators to drastically reduce production and increase costs, thereby destroying profits. The first thing they did was to rehire people we had let go in recent years for incompetence or stealing or for many other legitimate reasons, thereby enlarging the payroll beyond economic justification. They then decreed that miners could come out of the mine for lunch break. It takes considerable time for the men to get underground and to their workplaces, due to the small size of the cages in which they are lowered and the distances to their workplaces after they are underground. It is customary all over the world for miners to take their lunches underground with them. By letting the men come out for lunch they lost close to two hours of their productive time, so production dropped off very quickly and the mines lost money.

We immediately filed for reimbursement for the gold and other metals on hand and for the equipment and so forth, and we had lawyers pursuing the claims. It was several years after I retired before the companies received any reimbursement, but eventually they did get some compensation.

While we were negotiating with the government officials there were also negotiators from the large banks in the United States that had large outstanding loans. The Sandinistas said that they would not recognize any loans made to the previous regime, but everyone knew that the new regime had to get credit somewhere. What surprised me was that the bank representatives were practically begging the new government to take out more loans so that they could pay interest on existing loans so that the banks would not have to report defaults.

The main hotel in Managua was the Intercontinental, and the Sandanistas practically took it over. The restaurant was converted to buffet service only and one had to get in line to eat. The line was long and several times when the food was in sight some armed soldiers would push in front of me. The only prudent thing to do was step aside, but it certainly was aggravating. Often, there wasn't much left to eat when I finally arrived at the buffet.

In the early 1960s I was approached by a geologist that I knew who had worked at Frontino Gold Mines in Colombia before my time there. He was working as a consultant and had been asked by an English company that was interested in acquiring gold deposits to go to Costa Rica to examine an area of which they had knowledge. Since the area was primarily alluvial mining, and International Mining was interested in dredging properties, I agreed to go along. The geologist, Rocky, had arranged through a lawyer in San Jose for two natives to meet us near Sirena on the Peninsula de Osa and accompany us on our exploration of the area. Rocky had assured me that his Spanish was very good and that we would have no problem communicating on our trip through the jungle. We spent one evening with the lawyer in San Jose and then flew early the next morning to the Peninsula de Osa, where we landed on the beach at low tide near Sirena. The two natives soon appeared, and as soon as we started talking to them I knew that the stories I had heard about Rocky being something of a braggart and a know-it-all were quite true. His Spanish wasn't nearly as good as mine, which was very limited, but I was able, with very little help from Rocky, to arrange for our trip. They took us to their small farm where they had a few animals, thatched-roof shacks, their wives, and several small children. We laid out the food, utensils, hammocks, gold pans, shovels, and so forth and started loading up our backpacks in four equal proportions. Rocky with his usual bravado said, "Hell, Pat, I am a lot bigger than you are so give me a larger share." I agreed, even though I have always carried my fair share and each of us gave a few extra things to Rocky.

It was the rainy season and very muddy all around the area, including inside the shack, which had a dirt floor and no door. One of the women had fixed some soup and rice and the four of us sat around a very small table with pieces of logs as seats. The chickens, a monkey, and a dog were trying to get at our food while a pig was trying to knock us off our stools. The rice was good but the soup was hardly palatable, so we cut open some cans of beans, which we ate out of the cans. We slept on some planks in a nearby shed with our jackets as pillows and our raincoats as cover. Early the next morning we started out and walked along the beach to the Sirena River to see a washing plant operation that had worked for several years but was no longer operating. There had been a storm a few days before, and along the beach there were several men working shoveling black sands into small sluice boxes. The sands contained fine gold that had been reconcentrated by the recent high waves. It was what one could call a gold farm, as gold was frequently washed up during storms. The men were recovering enough fine gold for a rather meager existence and had been doing it for several years along the same beaches. We visited the old Sirena operation, which apparently had been reasonably productive, judging by the extensive areas of operations. We then set off on an old trail through the jungle towards the Rincon River where there were several small mining operations. It was raining and slippery underfoot, and Rocky slipped and fell soon after we started down a small slope. He hollered, "I offered to take a larger portion, but you fellows must have given me all the heavier things." So we took part of his load and distributed it among the three of us. The same thing happened again a few miles further on, where we were pelted by nuts thrown at us by monkeys overhead in the trees. Rocky was reduced to a very light load but still complained. We camped that night by a small stream and hung our canvas-covered hammocks on trees. I was looking forward to a night's rest under cover from the rain, but just before we turned in one of our guides told us about a man who had been killed in a similar hammock by a large boa constrictor that crushed him as it wound around the hammock. That spoiled my rest, worrying about getting killed, as there were many snakes in the jungles there. We visited several small mines in the next three days and panned for gold in many streams. On the third day the sun finally came out, but we became almost as wet from perspiration as from the rain. Late in the day we came upon a small pool where we stripped down and had a much-needed bath. By then I was either not so worried about being crushed by a boa or just too tired to worry, and I slept fairly well in the hammock.

During our journey we had seen several small mining operations that were making money and enough indications of gold where we had panned that I thought it would be possible to develop an area large enough to support a dredging operation. However, a drill crew that I sent in later prospected for several months but could not develop a large enough area of economically dredgeable ground.

I did not hear much about Rocky for several years, until I saw him in Mexico where he was examining a property for a client. Not long afterward I heard that he had left Durango one evening in a new four-wheel-drive vehicle en route to Mazatlan and had disappeared. There was extensive searching for him but no trace was found until two years later when two men were arrested on other charges and finally confessed that they had heard Rocky bragging in a bar in Durango about his new vehicle. They had left ahead of him, put in a road block, killed him when he stopped, buried him in a shallow grave close to the road, and gone off with his vehicle. His body was recovered and taken to Colorado for burial, where his wife was still living. It was an unnecessarily tragic end for a good geologist who was his own worst enemy.

In the early 1980s I consulted for a Canadian company in the Kalimantan area of Borneo. I flew to Jakarta, Indonesia, where I met a Chinese engineer who was manager of Pacific Tin in Malaysia. We met with the Indonesian engineer who had presented the property and went to the Mines Ministry to gather information on the area that we were to visit. We also spoke to officials there, who told us that the government would get 50 percent of any production and would not invest any money in exploration or development. We flew to a small town near the mouth of a river on the southern coast of Kalimantan and stayed at a very small hotel where I was introduced to the custom of eating with your fingers, as no utensils were provided. We arranged for a small open boat with an outboard engine to go up the river, and at the end of a very long day we reached about as far as we could go when the engine stopped. There was a very small village nearby with just a few shacks. The Indonesian engineer located the chief of the village, who said that we could stay in his home but that it would be crowded. The chief's wife gave each of us a bowl of rice, and ten of us, including the family, slept on the dirt floor without anything but the clothes we had on. It was so crowded it was hard to turn over, and the rats were crawling around and the bats were flying around so it was a most uncomfortable night.

The boat owner managed to repair the motor, and we spent three days prospecting, sleeping in the boat at night. We were primarily looking for dredgeable areas, and although we did see a few people panning on the river bars there was not enough area for dredging anywhere, and in view of the government's unreasonable demands we did not recommend going ahead with the project.

I had another interesting consulting job in Papua New Guinea for an Australian group. There had been a great deal of gold dredging in Papua New Guinea from the 1930s to the time of the Japanese invasion and then again after the war for a few years with dredges that were salvaged. The manager of Bulolo, the company that was operating there, and another engineer had buried the gold bars away from the camp as the Japanese were approaching. After the war, the manager had found the bars and used the proceeds to re-build two of the dredges, which operated for a few years. International Mining purchased one of those dredges later on and shipped it to Bolivia, where it was installed and operated for many years as I have mentioned elsewhere. The property that I went to see for the Australian group was on the other side of a mountain from the Bulolo operation. The values they were finding in the prospecting were marginal, not anywhere near as good as what the Bulolo operation had. Although the government would grant a concession or permit to mine a certain area, you had to work out an agreement with the local tribes as well. Different tribes controlled most of the area in that part of New Guinea and it was very difficult to reach an agreement with any of them on putting in an operation. Even in the area where materials would have to be brought in by boat, the tribe in that area made unreasonable demands for payment for peaceful travel through their land. I spoke at length with a member of the tribe that controlled the area that was being prospected. He said that despite efforts by the government and the missionaries the various tribes were still fighting each other, and when a member of one tribe was killed that tribe would not rest until they killed someone, and it didn't matter who it was. He said that was what had happened to Michael Rockefeller when he came ashore and encountered a tribe that was looking for someone to kill. In view of the low values, the problems with the tribes, and the lack of government support and financing, I did not recommend further work on the project.

My brother Bill was doing some consulting for a group in Sierra Leone in Africa and was in New York on his birthday in negotiations with the minister of mines from Sierra Leone. He had said that he would be out to New Canaan for a birthday dinner and Sandra had made a cake. He called and said that he was still meeting with the minister and had more to discuss and asked if it was all right if he brought the minister out, and we said certainly. Sandra served an excellent dinner, and we had a very pleasant and interesting time as the minister was Oxford educated and very interesting. Bill called later that night and said that the minister knew that Sandra was from the South, and he was most appreciative of the very gracious way she had entertained him. He approved the project that Bill had proposed.

We had been talking about visiting New Zealand for some time, and one midwinter vacation we decided it was time to stretch the budget, and we went on a tour there with a few days in Australia as well. It was a great trip, and the day after we returned I had a call from a company in Idaho asking me to visit a property on the coast of the South Island of New Zealand. I returned to New Zealand the next day and spent several days onsite and then some time at corporate headquarters in Idaho. Values were marginal and there were operating difficulties, so they did not go ahead on the project. Another group did take on the project a year or two later, but it was not very successful.

While we were at Rotorua in New Zealand Sandra heard that there was good fishing on the lake there, and she arranged for a boat and guide to pick us up at 5:00 a.m. We had a good fishing trip and caught large trout that the chef at the lodge prepared for us for lunch. They were excellent.

I have been most fortunate in having an active, interesting, and healthy long life, and I have been blessed with wonderful family and friends. For most of my life my last thoughts every night are thanking God for all the days of my life.